Springer Texts in Business and Economics

Springer Texts in Business and Economics (STBE) delivers high-quality instructional content for undergraduates and graduates in all areas of Business/Management Science and Economics. The series is comprised of self-contained books with a broad and comprehensive coverage that are suitable for class as well as for individual self-study. All texts are authored by established experts in their fields and offer a solid methodological background, often accompanied by problems and exercises.

Alex Coad

Data Science MBA

Big Data, Digitalization, and Strategy;
With Applications in R

Alex Coad
Waseda Business School
Waseda University
Tokyo, Japan

ISSN 2192-4333 ISSN 2192-4341 (electronic)
Springer Texts in Business and Economics
ISBN 978-981-95-2432-7 ISBN 978-981-95-2433-4 (eBook)
https://doi.org/10.1007/978-981-95-2433-4

© The Editor(s) (if applicable) and The Author(s), under exclusive license to Springer Nature Singapore Pte Ltd. 2025

This work is subject to copyright. All rights are solely and exclusively licensed by the Publisher, whether the whole or part of the material is concerned, specifically the rights of translation, reprinting, reuse of illustrations, recitation, broadcasting, reproduction on microfilms or in any other physical way, and transmission or information storage and retrieval, electronic adaptation, computer software, or by similar or dissimilar methodology now known or hereafter developed.
The use of general descriptive names, registered names, trademarks, service marks, etc. in this publication does not imply, even in the absence of a specific statement, that such names are exempt from the relevant protective laws and regulations and therefore free for general use.
The publisher, the authors and the editors are safe to assume that the advice and information in this book are believed to be true and accurate at the date of publication. Neither the publisher nor the authors or the editors give a warranty, expressed or implied, with respect to the material contained herein or for any errors or omissions that may have been made. The publisher remains neutral with regard to jurisdictional claims in published maps and institutional affiliations.

This Springer imprint is published by the registered company Springer Nature Singapore Pte Ltd.
The registered company address is: 152 Beach Road, #21-01/04 Gateway East, Singapore 189721, Singapore

If disposing of this product, please recycle the paper.

Preface

The digital age has brought with it a number of transformations such as new organizational structures, new business practices, new teamwork routines, and new ways of organizing work. Managers need to know how to work in data intensive organizations. Managers need not necessarily learn how to be skilful at writing code, however, or become highly proficient at advanced statistical analysis. The popular press often refers to the shortage of Data Scientists, but there is also a shortage of digitally-literate managers who possess the valuable "fusion skills" that combine strategic insights with data literacy. Managers need to be able to hold important conversations with data analysts. This book seeks to provide an overview of digital strategy considerations, as well as an overview of some of the most important concepts for data science and data analytics in business contexts. Some of the non-trivial concepts discussed in this book include overfitting, unsupervised learning, potential tradeoffs between predictive power and explainability, and the many types of ethical problems affecting data science and AI. This book is written with an audience of MBA students in mind: who are curious about important management concepts, who seek familiarity with data analysis, and who will need to be able to have conversations with data scientists and technicians (although they will not necessarily become data scientists themselves).

This book seeks to blend strategic aspects of digital transformation together with data analytics techniques. Discussions of digital themes such as ethical problems of AI bias, for example, become clearer when we work with data and see for ourselves how biases can persist even if we deliberately avoid including an explanatory role for sensitive variables. Working with actual examples helps highlight the tradeoffs between explainability and predictive power. Working with practice datasets can help managers develop the vocabulary they will need for discussions with their quantitative analysts.

There is a huge literature on data science and digital transformation, although it often focuses on specific topics (such as artificial intelligence, stages of digitalisation, and so on). This book tries to weave the pieces together to give a global view of research and practice in this area that could be useful in teaching an MBA course. This book came about in response to a perceived gap in the literature. Some early books on digital transformation proved to be valuable, but by now are a bit dated (Provost and Fawcett 2013; Schmarzo 2016; Kenett and Redman 2019). Schmarzo (2020) builds on the same ideas as Schmarzo (2016), but is less academic in tone. The work of Rogers (2016; 2023) is excellent regarding digital transformation and strategy, although it does not include data analytics, and the books do not focus on topics such as the ethics of data analytics and AI. Taddy (2019) is an important book, that covers the *Data Science MBA* course that Taddy taught at the University of Chicago. Since then, he left Chicago in 2018 to join Amazon as a vice president, although an updated and extended version of his 2019 book was published with the help of two coauthors (Taddy et al. 2023). Taddy's work is highly recommended, although it focuses on statistics and analytics techniques, whereas this book seeks to also cover themes relating to organizational transformation, digitalization, and business strategy that accompanies the application of data science thinking in business contexts. Some of the material in Taddy et al. 2023 (and also Taddy 2019), while extremely interesting, might be quite challenging for many MBA students. Relatedly, there are some excellent texts on statistical learning in R (James et al. 2021, Hastie et al. 2017) but they do not discuss the business strategy aspects. The book provides an overview of various techniques (such as Lasso regression, decision trees, and text analysis), but it is not a detailed textbook on these techniques. Instead, interested readers are given pointers regarding how to analyze topics in more depth, if this is what they want. At the end of each chapter, I make some suggestions for further reading. Regarding the data analysis techniques, I often refer the reader to Taddy et al. (2023) and James et al. (2021).

The practice of data analysis in this book is done using R software. R is a leading software package for statistical analysis, that has been specifically recommended by many data scientists, such as Schmarzo (2016); Kenett & Redman (2019); Taddy (2019); Taddy et al. (2023); James et al. (2021); Imai and Williams (2022); and Wickham et al. (2023). Python is another popular software language, which is particularly appropriate for tasks involving working with unstructured data such as text and images, and for deep learning. However, tasks that can be done in Python can generally also be done in R, and this is especially true for the scope of this book. In any case, industrial data scientists will be expected to be able to work with both R and Python, and learning R is a great place to start.

R code and datasets can be downloaded from my website https://alexcoad.com/ or from https://github.com/alexcoad.

Important terms appear in **bold** near their first mention. Terms in R appear in `consolas` font.

Tokyo, Japan   Alex Coad

References

Hastie, T., Tibshirani, R., Friedman, J. H. (2017). The Elements of Statistical Learning: Data Mining, Inference, and Prediction. Second Edition. New York: Springer.
Imai, K., & Williams, N. W. (2022). Quantitative Social Science: An Introduction in Tidyverse. Princeton University Press.
James, G., Witten, D., Hastie, T., & Tibshirani, R. (2021). An Introduction to Statistical Learning with Applications in R. 2nd Edition. Springer Nature, New York, NY, USA. Free to download from: https://www.statlearning.com/.
Kenett, R. S., & Redman, T. C. (2019). The Real Work of Data Science: Turning data into information, better decisions, and stronger organizations. John Wiley & Sons.
Provost, F., & Fawcett, T. (2013). Data Science for Business: What you need to know about data mining and data-analytic thinking. O'Reilly Media, Inc.
Rogers, D. L. (2016). The digital transformation playbook: Rethink your business for the digital age. Columbia University Press.
Rogers, D. L. (2023). The Digital Transformation Roadmap: Rebuild Your Organization for Continuous Change. Columbia University Press.
Schmarzo, B. (2016). Big Data MBA: Driving business strategies with data science. John Wiley & Sons.
Schmarzo, B. (2020). The Economics of Data, Analytics, and Digital Transformation: The theorems, laws, and empowerments to guide your organization's digital transformation. Packt Publishing Ltd.
Taddy, M. (2019). Business data science: Combining machine learning and economics to optimize, automate, and accelerate business decisions. McGraw Hill Professional.
Taddy M., Hendrix L., Harding M.C. (2023). Modern Business Analytics. Practical Data Science for Decision Making. McGraw Hill, New York, NY
Wickham H., Çetinkaya-Rundel M., Grolemund G. (2023). R for data science: import, tidy, transform, visualize, and model data (Second Edition). O'Reilly Media, Inc. Free to read online: https://r4ds.hadley.nz/

Contents

1 Introduction and Definitions 1
  1.1 The Current State of Digitalization 1
  1.2 Digitization and Digitalization 2
  1.3 The Three Vs of Big Data 6
  1.4 Data Science, Machine learning, Artificial Intelligence 7
  1.5 Lessons from History: Electrification 10
  1.6 Digital as a GPT 11
  1.7 The Productivity J-curve 12
  References 14

2 Digital Transformation of Organizations 17
  2.1 Models of Digital Transformation 17
     2.1.1 The Digital Transformation Playbook (Rogers 2016) 17
     2.1.2 Schmarzo's BDBMMI 21
     2.1.3 Iansiti and Lakhani (2020) 25
     2.1.4 Comparing these Models 27
  Further Reading 28
  References 28

3 Big Data Technologies and Architecture 31
  3.1 A Historical Perspective on Appropriate Organizational Structure 31
  3.2 A Single Data Lake 32
  3.3 Cloud 34
  3.4 APIs and Microservices 35
  3.5 Small "Agile" Teams 37
  3.6 Analytics Sandbox Environment 38
  3.7 R Example: Data from an API 40
  Further Reading 41
  References 41

4	Data Science in Organizations	43
	4.1 The Difference Between Data Science and Applied Statistics	43
	4.2 Skills of a Data Scientist	44
	4.3 Skills of Leaders for Digital Transformation	46
	4.4 Organizational Culture	49
	Further Reading	51
	References	51
5	**Statistical Associations Using Regression**	53
	5.1 Univariate Distributions	53
	5.2 Correlations	54
	5.3 OLS Regression	56
	5.4 Logistic Regression	58
	5.5 R Example: OLS and Logistic Regression	60
	5.6 Lasso Regression	63
	5.6.1 Overfitting	63
	5.6.2 An Introduction to Lasso Regression	65
	5.6.3 R Example: Lasso Regression	67
	5.7 Alternative Regression Models	70
	5.8 R Example: OLS Regression on Salary Data	72
	Further Reading	74
	References	74
6	**Ethics of Data Science and AI**	75
	6.1 Good News: Data can Elucidate how Decisions are Made	76
	6.2 AI Ethics	77
	6.2.1 AI Bias Due to the Training Data	78
	6.2.2 Black Box Algorithms and Explainability	79
	6.2.3 Privacy	81
	6.3 Competing Notions of Fairness	82
	6.4 Problems of Machines Lacking Accountability	83
	6.5 Expertise is Needed	86
	6.6 R Example: Logistic Regression on Loan Data	87
	Further Reading	88
	References	89
7	**Working with Data**	91
	7.1 Data Quality	91
	7.2 Common Problems of Working with Data	93
	7.3 Data Pre-processing	95
	7.4 R Example: PCA on Scoreboard Data	98
	Further Reading	102
	References	102

8 The User Experience (UX) 105
- 8.1 Customers are Spoilt 105
- 8.2 The Value of Customer Insights 106
- 8.3 UX and Innovation 107
- 8.4 Presenting Relevant Actionable Data with Dashboards 109
- Further Reading 111
- References 112

9 Data Visualization 113
- 9.1 The Importance of Graphs 113
- 9.2 Data Visualization Principles 113
  - 9.2.1 Basic Principles of Data Visualization 114
  - 9.2.2 Pre-attentive Processing 115
  - 9.2.3 Further Principles of Visual Design 116
- 9.3 Some Common Types of Graph 117
  - 9.3.1 Pie Charts 117
  - 9.3.2 3D Plots 118
  - 9.3.3 Overplotting and Some Remedies 119
  - 9.3.4 R Example: Binned Scatterplot 120
- 9.4 Communicating Graphs 121
- 9.5 Designing Dashboards 122
- Further Reading 123
- References 123

10 CART and Prediction 125
- 10.1 CART Models for Prediction 126
  - 10.1.1 Introduction to CART 126
  - 10.1.2 Decision Trees and the Variable Space 127
  - 10.1.3 CART Algorithm 127
  - 10.1.4 Greedy Algorithms 128
- 10.2 R Example: Predictive Power of CART vs OLS, on Bivariate Data 129
- 10.3 Random Forests 132
- 10.4 R Example: Lasso, CART and Random Forests on Real Estate Data 133
  - 10.4.1 Cleaning the Data 133
  - 10.4.2 Lasso 134
  - 10.4.3 CART 135
  - 10.4.4 Random Forests 136
- Further Reading 137
- References 137

11 Text as Data 139
- 11.1 The Bag of Words Model 140
  - 11.1.1 Tokenization 140
  - 11.1.2 Stopwords 141

	11.1.3	Stemming	143
	11.1.4	Document Term Matrix (DTM)	143
	11.1.5	Term Frequency and tf-idf, with an R Example	145
	11.1.6	Text Regression	146

11.2 Sentiment Analysis, with R Example 147
11.3 Topic Modelling 149
Further Reading 152
References 152

12 Causal Inference 153
12.1 Correlation Is Not Causation 154
12.2 Causal Language 156
12.3 Directed Acyclic Graphs (DAGs) 157
12.4 Techniques for Causal Inference 160
    12.4.1 Randomized Controlled Trials (RCTs) 160
    12.4.2 Natural Experiment 160
    12.4.3 Regression Discontinuity Design (RDD) 161
    12.4.4 Instrumental Variables (IV) 162
Further Reading 163
References 163

13 Concluding Remarks 165

Bibliography 167

Index 169

About the Author

Alex Coad is a Professor at Waseda Business School (Waseda University, Tokyo, Japan), and his research focuses mainly in the areas of firm growth, firm performance, entrepreneurship, and innovation policy. Alex has taught an MBA course on "Data Science for Management" at Waseda every year since 2020. Alex has published over 100 articles in international peer-reviewed journals. According to Google Scholar, Alex has over 17'000 citations and an H-index over 50. Alex is an Editor at the journals 'Research Policy' (on the Financial Times Top 50 list of journals for Business Schools) and 'Small Business Economics', and is an Associate Editor at 'Industrial and Corporate Change'. Previously Alex obtained a PhD from Université Paris 1 Panthéon-Sorbonne and the Sant'Anna School, Pisa, Italy, and held academic positions at the Max Planck Institute (Jena, Germany), Aalborg University (Denmark), SPRU (Univ. Sussex, UK), and CENTRUM Graduate Business School (Lima, Peru), and also being an Economic Analyst at the European Commission (IRI group, JRC-IPTS, Sevilla). In December 2016, Alex received the 2016 Nelson Prize at University of California Berkeley. In 2024, Alex co-authored (along with Anders Bornhäll, Sven-Olov Daunfeldt, and Alex McKelvie) the Open Access book entitled "Scale-ups and High-Growth Firms: Theory, Definitions, and Measurement", also published by Springer.

List of Figures

Fig. 1.1	Big Data, Data Science, Machine Learning, and Artificial Intelligence AI builds on ML, which builds on Data Science, which builds on Big Data. (*Source* Author's elaboration)	9
Fig. 1.2	Competitive dynamics: digital firms and traditional firms (*Source* Author's elaboration, similar in style to Iansiti and Lakhani [2020])	14
Fig. 2.1	Rethinking the marketing funnel in the digital age. (*Source* Author's elaboration, inspired by Rogers [2016])	19
Fig. 2.2	The Big Data Business Model Maturity Index (BDBMMI). (*Source* Author's elaboration, inspired by Schmarzo [2016])	22
Fig. 2.3	Four stages of digital operating model transformation. (*Source* Author's elaboration, inspired by Iansiti and Lakhani [2020, p. 119])	25
Fig. 3.1	Stock price dynamics for Nvidia. (*Source* Author's calculations using the quantmod package, see R code file for details)	40
Fig. 5.1	Univariate analysis: histogram and kernel density plot. (*Source* Author's elaboration. See R code)	54
Fig. 5.2	Bivariate clouds of datapoints and the corresponding correlation coefficients. (*Source* DenisBoigelot, original uploader was Imagecreator, CC0, via Wikimedia Commons. https://commons.wikimedia.org/wiki/File:Correlation_examples2.svg)	55
Fig. 5.3	OLS regression. The graph shows 5 raw datapoints (black circles), OLS line of best fit (orange line), OLS predicted values (orange triangles), and the error terms (blue dashed lines). (*Source* Author's elaboration, see R code for details)	57
Fig. 5.4	OLS regression output in R. (*Notes* Author's elaboration, see R code file for details)	57

Fig. 5.5	OLS line of best fit, for the relation between effort and test score. (*Notes* Author's elaboration, see R code file for details)	61
Fig. 5.6	OLS (left) and logistic regression (right) models for the case of the binary outcome variable. (*Notes* Author's elaboration, see R code file for details)	62
Fig. 5.7	A well-fitting model (centre) in between an underfit model (left) and an overfit model (right). (*Notes* Author's elaboration, see R code file for details)	64
Fig. 5.8	A rough intuition of how Lasso shrinks regression coefficients to zero. Left: unconstrained OLS model. An OLS regression might yield coefficient estimates that are normally distributed around zero. Right: corresponding histogram of regression estimates from Lasso. Arrows indicate Lasso shrinkage. Lasso shrinks all the coefficients towards zero, by an approximately equal amount, leading to a mass point of coefficients at a value of precisely zero. (*Source* Author's elaboration)	67
Fig. 5.9	Lasso path plot, for Scoreboard data. (*Source* Author's elaboration, see R code for details)	69
Fig. 5.10	gamlr output, for Lasso regression. (*Source* Author's elaboration, see R code for details)	70
Fig. 5.11	Lasso output: coefficients. (*Source* Author's elaboration, see R code for details)	70
Fig. 5.12	Lasso analysis: predicted values. (*Source* Author's elaboration, see R code for details)	71
Fig. 5.13	OLS regression, model 1. (*Source* Author's elaboration, see R code for details)	72
Fig. 5.14	Scatterplot of the relationship between Salary and Experience, with an OLS best-fit line overlaid. (*Source* Author's elaboration, see R code for details)	73
Fig. 5.15	OLS regression output for Models 2 and 3. (*Source* Author's elaboration, see R code for details)	73
Fig. 6.1	Prediction, Judgment, and Decisions: 2 regimes with different payoffs. (*Source* Our elaboration, drawing on ideas in Agrawal et al. [2022])	77
Fig. 6.2	How bias enters the AI loop, and how to cut the loop. (*Source* Author's elaboration)	79
Fig. 6.3	A trade-off between the predictive power of techniques, and their explainability. (*Source* Author's elaboration)	80
Fig. 7.1	Examples of data collection errors, including duplications (in orange), missing values (in green), and inconsistencies (in blue). (*Source* Author's elaboration)	93

Fig. 7.2	Scatterplot of total sales, and employment, with a contextual outlier that has high sales and low employment. (*Source* Author's elaboration, see the R code file for details)	94
Fig. 7.3	Data imputation, and the creation of an "imputed" dummy. (*Source* Author's elaboration, drawing on ideas in Taddy et al. [2023, p. 124])	97
Fig. 7.4	Summary statistics in R. (*Source* Author's elaboration, see R code for details)	99
Fig. 7.5	inspecting the 6 PCA components in R. (*Source* Author's elaboration, see R code for details)	99
Fig. 7.6	scatterplot matrix in R. (*Source* Author's elaboration, see R code for details)	100
Fig. 7.7	Missing values in the data. (*Source* Author's elaboration, see R code for details)	101
Fig. 7.8	Scatterplot matrix in R. (*Source* Author's elaboration, see R code for details)	102
Fig. 8.1	DX shortens the innovation cycle. (*Source* Author's elaboration, inspired by Rogers [2016])	109
Fig. 8.2	Example of a user-facing dashboard, for a laundry smartphone app. (*Source* Author's elaboration, inspired by https://laundry.senkaq.com/, with graphical help from mistral.ai)	111
Fig. 9.1	Common features for pre-attentive processing (Notes Author's elaboration, similar in style to Nussbaumer Knaflic [2015, Fig. 4.4] and Wexler et al. [2017, Fig. 1.10])	115
Fig. 9.2	The rule of thirds, and the Z-pattern of scanning (*red dashed arrows*) (*Source* Author's elaboration)	116
Fig. 9.3	Graphs in Microsoft Excel. A standard 3D pie chart on the left, and a horizontal bar chart showing the same information on the right (*Source* Author's elaboration)	118
Fig. 9.4	3-Dimensional plot in Microsoft Excel (*Source* Author's elaboration)	118
Fig. 9.5	Scatterplot in R (*Notes* Author's elaboration, see R code file for details)	119
Fig. 9.6	Both the jitter option with semi-transparent datapoints (*left*) and the contour plot (*right*) reveal the phenomenon of overplotting (*Notes* Author's elaboration, see R code file for details)	120
Fig. 9.7	Traditional scatterplot (*left*) and a binned scatterplot (*right*) (*Notes* Author's elaboration. See R code file for details)	121
Fig. 9.8	Dashboard example made from fictional data (*Source* Author's elaboration, in the style of Wexler et al. [2017])	123

Fig. 10.1	The Eurostat-OECD (2007) HGF definition, represented as a decision tree (left) and represented as a partitioned input space (right) (*Notes* Author's elaboration, drawing on Karlsson and Coad [2025])	128
Fig. 10.2	Recorded speech data, using the data and R code from Shumway and Stoffer (2025)	130
Fig. 10.3	OLS applied to a subset of the speech data (*Notes* Author's elaboration, see R code file for details)	131
Fig. 10.4	OLS and CART predicted values (*Notes* Author's elaboration, see R code file for details)	132
Fig. 10.5	How majority ruling in random forests leads to the classification of observations (*Notes* Inspired by Johan Karlsson, see Fig. 1.2 in Karlsson and Coad [2025])	133
Fig. 10.6	Lasso path plot (*Notes* Author's elaboration, see R code file for details)	135
Fig. 10.7	Trees fitted to the MLIT data. *Left*: univariate tree. *Right*: univariate tree that is a ridiculous case of overfitting (*Notes* Author's elaboration, see R code file for details)	136
Fig. 10.8	A more complete tree, fitted to the real estate data (*Notes* Author's elaboration. See R code file for details)	137
Fig. 10.9	Results of the Random Forest analysis of variable importance (*Notes* Author's elaboration. See R code file for details)	137
Fig. 11.1	Processing the raw text with tokenization and the removal of stopwords (*Notes* Author's elaboration, details in R code file)	143
Fig. 11.2	From raw text to the corresponding representation in a Document-Term Matrix (DTM) (*Source* Author's elaboration)	144
Fig. 11.3	Wordclouds: Adam Smith (*left*), Karl Marx (*centre*), and David Ricardo (*right*) (*Notes* Author's elaboration, see R code file for details)	146
Fig. 11.4	tf-idf statistics for Smith, Marx, and Ricardo (*Notes* Author's elaboration, see R code file for details)	146
Fig. 11.5	The most frequent negative-sentiment and positive-sentiment words in the novel "A Tale of Two Cities" by Charles Dickens (*Notes* Author's elaboration, see R code file for details)	148
Fig. 11.6	Wordcloud for the most negative and positive words (in terms of sentiment) in the novel "A Tale of Two Cities" by Charles Dickens (*Notes* Author's elaboration, see R code file for details)	149

Fig. 11.7	Sentiment analysis for Dickens, for 3 sentiment dictionaries (*Notes* Author's elaboration, see R code file for details)	150
Fig. 11.8	Adding a topic layer (*Source* Author's elaboration, in the style of Provost and Fawcett 2013, Fig. 10.6)	151
Fig. 12.1	If X and Y are correlated, this could mean five things (*Notes* Author's elaboration. *Arrows* denote the direction of causality)	155
Fig. 12.2	DAG representation of the example in Table 12.3	159
Fig. 12.3	*Left* The correlation between treatment and outcome cannot be given a causal interpretation, because of the existence of a confounder. *Right* Randomization cuts the arrow from confounder to the treatment; therefore any correlation between treatment and outcome can be given a causal interpretation Treatment → Outcome (*Source* Author's elaboration)	161
Fig. 12.4	DAG for understanding identification in Instrumental Variables (IV) analysis	162

List of Tables

Table 1.1	ICT usage by business populations, 2023 or latest available year	3
Table 1.2	Definitions of Digitization and Digitalization	5
Table 2.1	5 behaviours and strategies in the customer network	19
Table 3.1	How microservices enable the shift from Monolithic to Modular IT	37
Table 3.2	Comparing the data lake with the analytics sandbox environment	39
Table 5.1	Examples of values for the probability, odds, and log odds	60
Table 5.2	Alternative regression models	71
Table 6.1	Various notions of fairness, with definitions and examples	84
Table 6.2	A confusion matrix, which is a contingency table of observed and predicted cases	85
Table 6.3	Varying degrees of human involvement in machine learning processes	85
Table 6.4	Logistic regression on the loan data: coefficients and z statistics	88
Table 12.1	Language of associations vs language of causality	156
Table 12.2	Beer drinking and driving ability example	158
Table 12.3	Example of Simpson's paradox: medical treatment and heart attacks	159

Introduction and Definitions

The internet is already 50 years old. While some firms have thrived in the digital age, many others continue to lag behind. A major reason for the uneven level of digital maturity of firms is that digitization and digitalization affect strategy and organizational structure in non-obvious ways. Digital transformation (DX) requires a variety of skills: not just software programming skills, but a broader range of skills (Davenport and Harris 2017). Programming itself may even become less important in future, as artificial intelligence (AI) tools such as Copilot can provide assistance with programming. Executives need a common understanding of digital and AI, otherwise they will not be able to keep up with important strategic discussions. Executives will need to know the fundamental concepts of digital transformation and AI, although they will not need to have cutting-edge specialist expertise in technical tasks (such as building deep neural networks from scratch). This chapter provides a foundation for the rest of the book by presenting evidence on the state of digitalization, and discussing some key concepts related to big data and digital transformation.

1.1 The Current State of Digitalization

Table 1.1 shows some statistics on ICT usage by business populations, using the latest available data from the OECD. The first column shows that the share of employees that regularly use a computer in their work is less than two thirds in all countries. The second column shows that almost all businesses have a website or home page: the proportion is around 95% for OECD countries, and around 85% for Brazil. However, the number is lower when it comes to businesses with a website that allows for online orders or reservations. In general, such e-commerce capabilities are more common among large firms (with 250 or more employees) than for the broader population of firms with 10 or more employees. The share of enterprises having performed big data analysis varies by country, from a maximum of around 75% for the population of large firms in South Korea, Germany, and Italy, to values closer to 25% for Canada, Brazil, and Japan (although different definitions may hinder accurate comparisons across countries). Large firms are always more likely to have engaged in big data analysis than the broader population that includes small firms. Similarly, the use of AI differs across countries, with large firms being more active in terms of AI than small firms. Finally, the last column shows that most firms use social media, with large firms being more likely than small firms to use social media.

In sum, there is considerable variation across countries regarding ICT usage and digitalization. Most firms, but not all, seem to have a website, and are active on social media. There are still many employees who do not regularly use a computer in their work. There is considerable variation across countries in terms of the numbers of firms that have performed big data analysis, and only a minority in all countries have used AI. We can also consider that, just because a firm is active in a particular digital area, it does not mean that its efforts have been successful. There may be many firms that use computers, have websites, engage in e-commerce, and invest in AI, that are nowhere near being able to reap the full benefits of their digital investments. This book seeks to encourage firms to increase their investments in their digital transformations in order to reap the many benefits.

1.2 Digitization and Digitalization

Digitization refers to the process of converting information into digital format. This information may previously have been unrecorded, or may have been recorded in analogue format (such as writing comments by hand into a book of guest reviews). Digitization is the first step of the process that leads to the collection of big data, analysis using data science methods, and perhaps even machine learning and AI.

Table 1.2 contrasts the definition of Digitization with that of Digitalization.

Digitization can lead to profound and non-obvious transformations of the business context. At a first stage, one might not care too much if the song is stored in LP format or MP3 format, as long as we can listen to the same song. However, storing the same song in a digital format leads to knock-on effects that affect the organization of industry, and leads to the emergence of new business models. Digital music is downloadable, which replaces the need for physical music stores with the emergence of online platforms such as Napster and iTunes. This then leads to a shift in "connectivity" (Adner et al. 2019), because online platforms facilitate the sharing of music files, leading to the need for new rules regarding behaviour and legality. Connectivity also enables interactions between consumers regarding their preferences for songs, leading to a growing role for recommendations in the context of social media interactions, and also the emergence of recommendation

1.2 Digitization and Digitalization

Table 1.1 ICT usage by business populations, 2023 or latest available year

	Persons employed regularly using a computer in their work		Businesses with a website or home page		Businesses with a website allowing for online ordering or reservation or booking (e.g. shopping cart)		Businesses having performed big data analysis		Businesses using artificial intelligence (AI)		Businesses using social media	
	2019: Percentage of employment		2023: Percentage of enterprises		2023: Percentage of enterprises		2023: Percentage of enterprises		2023: Percentage of enterprises		2023: Percentage of enterprises	
	10+ Empl.	250+ Empl.	10+ Empl.	250+ Empl.	10+ Empl.	250+ Empl.	10+ Empl.	250+ Empl.	10+ Empl.	250+ Empl.	10+ Empl.	250+ Empl.
Australia			80.9	93.02	28.97	35.81	D 9.46	D 39.50	3.39	11.03	68.37	U 75.37
Canada			84.30	96.00	D 61.60	D 77.80	2.60	24.10	D 4.70	D 26.50	D 66.40	D 96.00
France			68.84	94.39	20.95	38.45	33.90	71.99	5.88	20.94	66.82	90.02
Germany			92.85	98.68	25.72	33.74	37.0799	74.59	11.55	35.39	60.68	87.73
Italy	55.85	56.65	74.22	92.19	16.25	31.17	26.6056	74.132	5.05	24.08	57.31	81.38
Japan				D 96.50			D 4.79	D 19.28	D 3.52	D 21.70		
Korea	65.79	58.00	70.19	92.79	38.46	34.91	39.65	51.18	27.99	36.42		
Spain	58.94	60.96	77.35	96.71	18.99	33.98	38.01	76.20	9.18	39.66	63.68	92.58
United Kingdom	64.40	65.28	83.39	95.56	24.50	34.88	26.87	49.86	4.11	11.48	71.75	84.05

(continued)

Table 1.1 (continued)

	Persons employed regularly using a computer in their work		Businesses with a website or home page		Businesses with a website allowing for online ordering or reservation or booking (e.g. shopping cart)		Businesses having performed big data analysis		Businesses using artificial intelligence (AI)		Businesses using social media	
	2019: Percentage of employment		2023: Percentage of enterprises		2023: Percentage of enterprises		2023: Percentage of enterprises		2023: Percentage of enterprises		2023: Percentage of enterprises	
	10+ Empl.	250+ Empl.	10+ Empl.	250+ Empl.	10+ Empl.	250+ Empl.	10+ Empl.	250+ Empl.	10+ Empl.	250+ Empl.	10+ Empl.	250+ Empl.
European Union (27 countries)	54.33	58.15	78.10	94.67	22.23	33.19	33.17	71.66	8.03	30.40	60.95	86.00
Brazil	49.00	40.00	55.36	84.65			6.72	24.03	12.93	40.78	88.23	92.03

Source Author's elaboration, based on https://data-explorer.oecd.org/ [last accessed 16th Oct 2024]

Notes: Various indicators of ICT usage by businesses, for those businesses with 10+ employees, or 250+ employees

D: means "Observation status: Definition differs"; and U: means "Observation status: Low Reliability"

Years used in the table columns:

Website: data refer to the most recent available year for Australia (2022), Canada (2021) Japan (2022), Korea (2022), and UK (2020)

ONLINE ORDERING: data refer to the most recent available year for Australia (2022), Canada (2021), Japan (2021), Korea (2022), and UK (2020)

BIG DATA: data refer to the most recent available year for Australia (2022), Canada (2021) Japan (2022), Korea (2022), and UK (2019)

AI: data refer to the most recent available year for Australia (2022), Canada (2021), Japan (2021), Korea (2022), and UK (2020)

Social media: data refer to the most recent available year for Australia (2022), Canada (2021), and UK (2019)

For details, refer to https://data-explorer.oecd.org/

1.2 Digitization and Digitalization

Table 1.2 Definitions of Digitization and Digitalization

	Digitization	Digitalization
Definition	The conversion of analogue data and processes into a machine-readable format	The use of digital technologies and data as well as the interconnection that results in new activities, or changes to existing activities *Digital transformation*: the economic and societal effects of digitization and digitalization
Example	The same song can be stored as a vinyl/LP record, or in MP3 format	The emergence of new businesses such as the iTunes store and Spotify

Source Author's elaboration, inspired by OECD (2019)

algorithms (e.g. on Spotify) and the shift from requested content to suggested content. Given that recommendation algorithms work better if they can draw on large datasets, the optimal industry structure could converge around a small number of large platforms.

The case of the music industry also helps to highlight the dangers of inaction in the face of the emergence of new digital technologies (Rogers 2016). 1993 saw the emergence of a new technical standard for compressing audiofiles: the MP3 format. MP3 enabled the compression of music into small digital files, with minimal loss of audio quality. 1993 also saw the emergence of Mosaic, the first popular Web browser. Putting these two together gave a powerful new value proposition for consumers: the ability to download and share music instantly and store it on their computer's memory. This new value proposition required a reaction from the industry incumbents, such as the RIAA (Recording Industry Association of America). One option could have been that the RIAA develop a new business model to better serve these customer needs using the new technology. The other option, that RIAA unfortunately decided to pursue, was to do nothing except for suing the companies creating the first portable devices for the storage and playback of MP3 audio files (Rogers 2016). Over the period 1999–2012, the recorded music industry saw its global sales drop from $28bn to $16bn. Meanwhile, Napster was launched in 1999 as an online platform for the illegal peer-to-peer sharing of MP3. RIAA missed the opportunity to develop a product offering that applied new technologies to meet customer needs, and it took an outsider (Apple, which was not an incumbent in the recorded music industry) to set up the iTunes store as a solution: a legal MP3 store that respects copyright considerations. Through its inaction, RIAA missed out on foregone revenue, and let itself be overtaken by an outsider (Apple). Furthermore, "[b]y waiting as long as possible to adapt what it offered to customers, the music industry trained millions of young listeners to expect digital music to be free" (Rogers 2016, p. 167).

Another example of how digitization leads to unexpected changes comes from the digitization of photography. Digital photos are cheaper to take and store, encouraging users to take more photos, which can then (due to their digital nature) be easily uploaded and shared on social media, where large databases are accumulated, thereby facilitating the training of machine learning algorithms and AI applications, further enhancing the customer experience regarding interaction via social media. Digital photos on smartphones thus enabled a new value proposition for customers: the effortless sharing of moments and experiences, fostering connection and community. Iansiti and Lakhani (2020, p. 7) go as far as to say this: "Ultimately, Kodak was not killed by Fuji or by a digital photograph startup, but by the emergence of smartphone and social network firms."

Digitization is therefore a first step along the DX path that can lead to unexpected new opportunities. Digitization can even be considered to be a necessary step on the ladder to the promised land of AGI (Artificial General Intelligence). In the previous era, different functions required their own stand-alone devices. Each function had its own separate hardware: audio used LP "vinyl" records and cassette tapes, video used videocassettes, clocks and metronomes were embodied in their own hardware, calculators came as separate pieces of electronic hardware, and so on. Digitalization allowed music and video to be stored as digital files on hardware that can be repurposed for other functions too, such as providing functions as metronomes and calculators. Now, all these functions are digital and can be combined onto the same device. Furthermore, computing power can be shifted from a personal computer to the cloud. Further steps up the "ladder of generality", taking us closer to AGI (Narayanan and Kapoor 2024) mean that different functions no longer need to have separate software, and can eventually be run by the fine-tuning of existing pre-trained models, perhaps even avoiding the need for programming by having instruction-tuned models where tasks are communicated verbally.

1.3 The Three Vs of Big Data

What is **Big Data**? How big does the dataset have to be, to be called Big Data? Is the amount of data the only thing that distinguishes between a big data strategy and a "normal" data strategy?

There is lots of hype and confusion about Big Data, which unfortunately some may find so off-putting that they would rather avoid any discussion of it at all (Davenport 2014). It has even been said that big data is like teenage sex[1]:

---

[1] Dan Ariely, Duke University; https://hewlett.org/is-big-data-like-teenage-sex/ [last accessed 27 Sept 2024].

> *Everyone talks about it; Nobody really knows how to do it; Everyone thinks everyone else is doing it; So everyone claims they are doing it...*

We begin with a definition from Davenport (2014, p. 1):

> "Big data refers to data that is too big to fit on a single server, too unstructured to fit into a row-and-column database, or too continuously flowing to fit into a static data warehouse."

Big data refers to data that has become too large and unstructured to behave according to traditional data analysis. More important than size per se, is the lack of structure in the data that poses particular challenges.

Discussions of big data often refer to the "three Vs":

- **VOLUME**: the amount of data that is created (the classic idea of how "big" the dataset is).
- **VELOCITY**: the speed at which new data are generated, as well as the speed at which they move around.
- **VARIETY**: the various types of variables, data files, and sources that create the data

More recently, three more Vs have been put forward (e.g. Lu 2020)

- **VERACITY**: trustworthiness of the data. With many forms of big data, quality and accuracy are difficult to control. This links in to the crucial task of data cleaning and preparation, that is discussed in Chap. 4.
- **VALUE**: considerations regarding how the data can actually be turned into value
- **VISUALIZATION**: the important task of understanding and communicating insights from the data, which will be discussed in Chap. 8.

Big data therefore poses many challenges to firms that are embarking on the digital transformation journey.

1.4 Data Science, Machine learning, Artificial Intelligence

A successful Digital Transformation (DX) starts with Big Data. Big data is not in itself a source of competitive advantage, although it is a key input. Many organizations have collected big data from various sources (customer records, sensor data from smart devices, email data, video data, social media data, etc.). However, not all the collected data actually gets analyzed. Of course, if the data does not get analyzed, it is hard to see how it could add value.[2] With the multiplication of data, the binding constraint is managerial attention (Adner et al. 2019).

Big data, by itself, could drown us in a deluge of information that paralyzes our decision-making abilities. Big data needs to be combined with data science, machine learning, and perhaps even AI. The main constraint is human attention. Data by itself is not useful unless it can be transformed into understandable information, and actionable knowledge, and wisdom to understand the broader context. For example, a QR code, in itself, is data that is hard to decipher with the human eye, but with the help of a QR reader it can (from the perspective of humans) be converted into new information (e.g. a website) that leads to an understanding of what actions to take (knowledge) as well as perhaps insights into how to reshape one's broader strategic direction (wisdom). Data science, ML and AI seek to help humans thrive in a world of big data by processing the data to give us relevant information leading to better decisions, while automating many low-stakes decisions using AI.

The age of big data emerged in the late 2000s, as firms sought to establish the necessary corporate infrastructure to collect the data in one place, to be able to better query it (HBR 2023). The age of Analytics started around 2009, as firms started counting things to answer questions having business value or product value. Progress in analytics came from new software that became more accessible and affordable. Data science builds on analytics to include more sophisticated methods for the analysis of data to yield useful models and predictions. Machine Learning (ML) refers to models that incorporate feedback loops to learn from the data about the best way to represent patterns in the data. Artificial Intelligence (AI) refers to applications that build upon Data Science and ML to prepare (and sometimes execute) data-driven decision-making. The relations between Big Data, Data Science, ML, and AI are depicted in Fig. 1.1.

Data science refers to the advanced skills and competencies that analyze Big Data resources. The goal of data science is to uncover new insights into the metrics and relationships between variables that help to predict and drive superior performance. "'Data Scientist' means a professional who uses scientific methods to liberate and create meaning from raw data."[3] Data science is similar to applied statistics, although there are some differences (Kenett and Redman 2019). Applied statisticians often focus on the mathematical properties of statistical estimators, and their formal proofs, while testing given hypotheses on well-behaved pedagogical datasets. Data scientists have a more holistic role, thinking carefully about which hypotheses could be interesting and relevant, collecting and cleaning the

---

[2] Merely having the data is not valuable per se. Many organizations have stores of data files that they have not yet looked at. For example, the US military routinely collected video footage from its drones, which was not analyzed due to a lack of human analysts (Davenport 2014, p. 19). Starbucks amassed huge amounts of data about customer transactions on loyalty cards, without having a clear idea regarding what to do with the data (Schmarzo 2016, p. 137).

[3] http://www.datascienceassn.org/code-of-conduct.html (Data Science Association 2018; last accessed 16th Oct 2024).

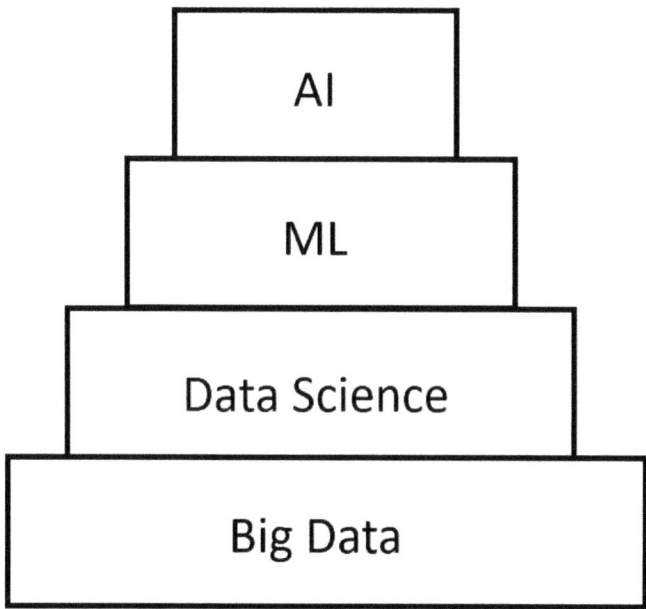

Fig. 1.1 Big Data, Data Science, Machine Learning, and Artificial Intelligence AI builds on ML, which builds on Data Science, which builds on Big Data. (*Source* Author's elaboration)

data themselves, embracing multidisciplinarity by drawing upon a broad set of relevant analytical methods (including qualitative insights from conversations with relevant stakeholders), and taking pains to clearly communicate their findings in non-technical terms to decision makers.

Machine Learning goes beyond Data Science because it involves more sophisticated techniques for learning from data, that incorporate feedback loops to converge upon interesting representations of patterns in the data. Machine Learning seeks to automate and scale the data science process. Two well-known styles of ML are **supervised** ML and **unsupervised** ML (Alpaydin 2021). The term "supervised" does not refer to human supervision, but the characteristics of the analytical task. Supervised learning is when the outcomes are labelled (e.g. "good" vs "bad"; picture of dog vs picture of cat), and the task is to find which variables help to predict a certain outcome. Unsupervised learning is when there is no specific outcome to predict, but the task is to find groups of similar entities by discovering patterns in how the datapoints are clustered together (e.g. which products are often found together in the same shopping basket).

Artificial Intelligence (AI) draws upon insights from ML to suggest and make decisions, thereby interacting with the world. A precise definition of AI is difficult, because there are many types of AI (Narayanan and Kapoor 2024).

1.5 Lessons from History: Electrification

In the late 1800s, there were concerns about an emerging technology: electricity. Popular fears included views that electricity was an unrestrained demon that could not only steal jobs but literally bring death by electrocution. Mary Shelley's novel "Frankenstein" owes part of its success to the fact that it resonated with public worries about electricity (Coeckelberg 2020). These days, fears surround data and AI, although many of these fears are exaggerated.

The acronym CEO used to correspond to "Chief Electricity Officer" (Vaz 2021, p. 3). These days, it usually means "Chief Executive Officer". Indeed, the use of electricity is now so pervasive that it would no longer seem relevant to have a role for a "Chief Electricity Officer", because we take electricity for granted. Perhaps soon the title "Chief Data Officer" will soon disappear, as claims about the importance of data for business performance become mundane.

The electrification of industrial factories is often mentioned in discussions of digital transformation, because electrification provides a useful parallel for thinking about achieving the full potential of big data in industrial applications (e.g. Brynjolfsson and McAfee 2014; Schmarzo 2016; Agarwal et al. 2022).

Before the arrival of electricity, steam-powered factories had an idiosyncratic layout. Power from the steam turbine was transmitted through a large central axle. This axle was preferably short, because long axles were more likely to break. As a consequence, the machines tended to be clustered around the short axle, with the most power-hungry machines occupying the positions nearest the axle. The need for proximity to the axle also meant that machinery was placed on the floors above and below the central steam engines.

When electricity arrived, the first reflex was to swap out the steam engine for the electric motor, and leave the rest of the factory layout unchanged. Under this arrangement, machines were positioned as close as possible to the energy source, which was a requirement under steam power but was not necessary with electric power. Under this arrangement, productivity did not increase by much.

Later on, factory designers realized that they were no longer bound by the constraints inherited from the steam power age. Factory layouts changed, although it took about thirty years (Brynjolfsson and McAfee 2014, p. 102). With the new factory layouts, each machine operated with its own small electric motor. In this way, the production process was based on the natural workflow of materials. Furthermore, electrification was followed by subsequent complementary innovations, such as lean manufacturing, Total Quality Management (TQM) and the 6 sigma approach. As a result, productivity doubled or tripled.

Similarly, factories in the computer age needed to be redesigned to benefit from new technologies. Nobel laureate Robert Solow famously quipped in 1987 that "You can see the computer age everywhere but in the productivity statistics." Unlocking the productivity gains from computer-assisted production required redesigning a broad range of activities including the scrapping the piece-rate pay system, allocating workers the authority for scheduling machines, changes in decision rights (because specialist workers may have a knowledge advantage), process and workflow innovation, more frequent and richer interactions with customers and suppliers, increased teamwork, increased lateral communication, as well as various other changes in skills, processes, structure, and culture (Brynjolfsson and Hitt 2000).

When thinking about productivity growth brought about by electrification, and digitalization, it can be helpful to distinguish between point solutions and system solutions (Agarwal et al. 2022).

Point solutions refer to incremental improvements at the task level, when one procedure can be swapped out and replaced with a new procedure, without changing the surrounding system. A point solution corresponds to a steam-powered factory, where the steam turbine has been swapped out and replaced with an electric turbine, but nothing else has changed. In this case, the productivity gains were small. Point solutions often focus on cost reductions in narrow tasks (including the replacement of humans with machines in narrowly-defined tasks). Point solutions are unable to envisage the new possibilities afforded by the potentially revolutionary new technology.

System solutions focus not on changing individual tasks but redesigning the production process from scratch, thereby improving existing procedures or allowing the introduction of new procedures. In the case of steam power, system solutions become possible when we throw off the constraints that affected the previous generation of technology (i.e. steam power, and its requirement for proximity to the central axle) and redesigning the factory based on the needs of the production line and the new possibilities afforded by electricity. System solutions focus not on cost reduction but on value creation and new growth opportunities brought about by dramatic productivity growth as well as the possibility of new product categories.

In sum, these lessons from history highlight the importance of understanding the broader context and implications of digital transformation, as well as the phenomenon of general purpose technologies (GPTs) such as electricity, and their role in shaping economic and social change.

1.6   Digital as a GPT

Innovation scholars have put forward the useful concept of **General Purpose Technologies (GPTs)** to describe a category of innovations with the potential to have huge impacts on economies and societies with far-reaching implications (Rosenberg 1982; Bresnahan and Trajtenberg 1995).

GPTs have three characteristics. First, they are pervasive because they are not restricted to particular sectors or applications. Second, GPTs are not born fully-formed, but they open up new technological trajectories that are the locus of ongoing technical improvement. Third, they enable various kinds of complementary innovations in many application areas.

Examples of GPTs include the steam engine, electricity, computers, and also Artificial Intelligence in the form of Generative Pre-trained Transformers (also known by the acronym GPT; as popularized by OpenAI's "ChatGPT" chatbot).

The value of GPTs is hard to ascertain when they first appear. GPTs improve in their technical characteristics, as well as triggering follow-on innovations in a variety of contexts, gathering momentum to eventually have a huge impact on productivity. It takes time to fully redesign the system to achieve all the GPT's benefits. Diffusion of a GPT's benefits takes time: starting perhaps with simpler point solutions, before reaching its full potential though broad-based redesign of systems in the form of systems solutions (as discussed in Sect. 1.5 in the case of the transition from steam power to electricity).

"The world's most valuable resource is no longer oil, but data" – read the title of a landmark article in the Economist magazine in 2017.[4] However, data is not intrinsically useful, per se, unless it is fully integrated into the system; i.e. until the productive system is designed around the power of data. Oil was a valuable GPT in its era, but oil per se is of limited value unless it is integrated in the system (Kenett and Redman 2019). Giving a jerrycan of petrol to a commuter using a horse and buggy is not particularly useful. Only if the person owns a car, and if there is complementary infrastructure (roads, traffic lights, parking spaces, highway code, car insurance, etc.) will petrol be useful for someone's commute. The challenge for firms in the era of data analytics as a GPT will be to build their businesses around their data assets to truly realize the productivity gains that are afforded by data resources.

1.7 The Productivity J-curve

The concept of a General Purpose Technology (GPT) has led to suggestions of a **productivity J curve** (Brynjolfsson et al. 2021), according to which the productivity gains of a new technology may be slow and unimpressive at first, but rapidly accelerating thereafter (Vaz 2021).

At the macro-economic level, the concept of a productivity J-curve helps to understand why the productivity gains from electricity (as an alternative to steam power) took decades to materialize. While computers have a broad potential for productivity growth, the productivity gains were not visible until after the adjustment costs were paid, and computers became fully integrated into production processes. Computers changed the workplace in many ways, changing the incentive systems (the elimination of piece-rate pay), as well as decision rights, business processes and workflows, the frequency and intensity of interactions with customers and suppliers, communication channels and teamwork practices, skills and training, and so on (Brynjolfsson and Hitt 2000).

---

[4] https://www.economist.com/leaders/2017/05/06/the-worlds-most-valuable-resource-is-no-longer-oil-but-data [last accessed 28th September 2024].

1.7 The Productivity J-curve

At the firm-level, the productivity J curve emphasizes that firms must make large up-front investments in digital technologies before reaping the benefits. To move beyond point solutions to benefit from system solutions, established incumbents will need to upgrade their infrastructure and redesign their business practices. Furthermore, the prevalence of intangible capital in our modern economies has changed the game for entrepreneurial firms, front-loading these firms with higher fixed costs, alongside lower variable costs (De Ridder 2024). Firms that can overcome the fixed costs will be well-positioned to rapidly scale up production as a consequence of the low variable costs (Coad et al. 2024). An example of the large fixed costs of initially investing in intangible assets comes from the case of iRobot, the producer of autonomous vacuum cleaners:

> "After I left academe in 2014, I joined the technical organization at iRobot. I quickly learned how challenging it is to build deliberative robotic systems exposed to millions of individual homes. In contrast, the research results presented in papers (including mine) were mostly linked to a handful of environments that served as a proof of concept." (Alexander Kleiner, quoted in Brynjolfsson et al. 2021, p. 333)

Another source of the J-curve effect for firms comes from network externalities. Digital firms often have business models that are aligned with the logic of network externalities: the greater the number of users, the greater the value to other users, the larger the training data, the better the learning algorithms, and the larger the ecosystem of platform stakeholders (such as sellers and app developers), and so on. Until the network reaches a critical mass, it can be difficult to recruit new participants to the network.

Iansiti and Lakhani (2020) highlight how digital firms compete with traditional firms in Fig. 1.2. Firms with digital operating models have a slow start, and may initially seem irrelevant for incumbents who would be able to create more value from the same number of users. Traditional firms may initially wonder whether digital firms will ever be able to catch up. However, while traditional firms suffer from decreasing returns to scale, digital firms who have overcome the initial difficulties can scale up at low cost and overwhelm the competition. Once the digital rival has finally caught up with a traditional firm, it is too late: the traditional firm stands no chance.

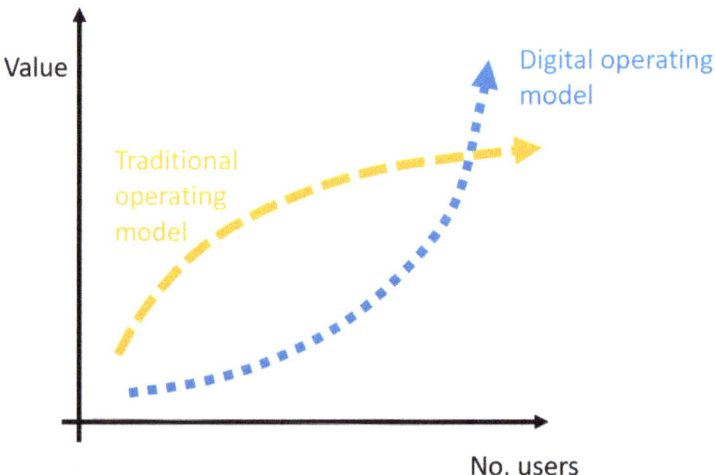

Fig. 1.2 Competitive dynamics: digital firms and traditional firms (*Source* Author's elaboration, similar in style to Iansiti and Lakhani [2020])

References

Adner, R., Puranam, P., & Zhu, F. (2019). What is different about digital strategy? From quantitative to qualitative change. Strategy Science, 4(4), 253–261.
Agrawal A., Gans J., Goldfarb A., (2022). Power and Prediction. Harvard Business Review Press, Boston, Massachusetts.
Alpaydin, E. (2021). Machine learning. Revised and Updated Edition. MIT Press Essential Knowledge series, MIT Press.
Bresnahan, T. F., Trajtenberg, M., (1995). General purpose technologies: 'engines of growth'? Journal of Econometrics 65(1), 83–108.
Brynjolfsson, E, Daniel, R., Chad, S. (2021). The Productivity J-Curve: How Intangibles Complement General Purpose Technologies. American Economic Journal: Macroeconomics, 13(1), 333–72.
Brynjolfsson E., Hitt L. M., (2000). Beyond Computation: Information Technology, Organizational Transformation and Business Performance. Journal of Economic Perspectives 14(4), 23–48.
Brynjolfsson, E., & McAfee, A. (2014). The second machine age: Work, progress, and prosperity in a time of brilliant technologies. WW Norton & Company.
Coad A., Bornhäll A., Daunfeldt S.-O., McKelvie A., (2024). Scale-ups and High-Growth Firms: Theory, Definitions, and Measurement. Springer Briefs: Springer Nature, Singapore, SG. Open Access. https://doi.org/10.1007/978-981-97-1379-0
Coeckelbergh, M. (2020). AI ethics. MIT Press. Cambridge, MA: USA.
Davenport, T. (2014). Big data at work: dispelling the myths, uncovering the opportunities. Harvard Business Review Press: Cambridge, MA.
Davenport T., Harris J. (2017). Competing on Analytics: The New Science of Winning. Harvard Business Review Press, Boston, MA, USA.
De Ridder, M. (2024). Market power and innovation in the intangible economy. American Economic Review, 114(1), 199–251.

References

HBR. (2023). HBR Guide to AI Basics for Managers. Harvard Business Review Press. Massachusetts: USA.

Iansiti, M., & Lakhani, K. R. (2020). Competing in the age of AI: Strategy and leadership when algorithms and networks run the world. Harvard Business Press.

Kenett, R. S., & Redman, T. C. (2019). The Real Work of Data Science: Turning data into information, better decisions, and stronger organizations. John Wiley & Sons.

Lu, J. (2020). Data Analytics Research-Informed Teaching in a Digital Technologies Curriculum. INFORMS Transactions on Education, 20(2), 57–72.

Narayanan, A., & Kapoor, S. (2024). AI Snake Oil: What Artificial Intelligence Can Do, What It Can't, and How to Tell the Difference. Princeton University Press.

OECD. (2019). Going Digital: Shaping Policies, Improving Lives, OECD Publishing, Paris, https://doi.org/10.1787/9789264312012-en

Rogers, D. L. (2016). The digital transformation playbook: Rethink your business for the digital age. Columbia University Press.

Rosenberg, N. (1982). Inside the Black Box: Technology and Economics. Cambridge University Press: Cambridge, UK.

Schmarzo, B. (2016). Big Data MBA: Driving business strategies with data science. John Wiley & Sons.

Vaz, N. (2021). Digital business transformation: How established companies sustain competitive advantage from now to next. John Wiley & Sons.

Digital Transformation of Organizations

2.1 Models of Digital Transformation

This chapter looks at three models of Digital Transformation. The first model (Rogers 2016) presents 5 key dimensions of DX (i.e. Customers, Competition, Data, Innovation, and Value), but does not put them into a sequential order. The second model (Schmarzo 2016) has five stages in a sequential order, and the third model (Iansiti and Lakhani 2020) has four stages in a sequential order. We examine each model sequentially, then compare their similarities and differences.

2.1.1 The Digital Transformation Playbook (Rogers 2016)

Rogers (2016) provides a useful model that covers 5 domains that have been picked up by much of the following DX literature:

- Customers
- Competition
- Data
- Innovation
- Value

These five dimensions will be briefly presented in the next subsections. These five are inter-related but are not necessarily placed in a causal order. As such, while we know which areas will require DX efforts, the temporal ordering of which steps to take might not be so clear (and are discussed later in Rogers' (2023) "Roadmap").

A general theme is that strategy matters more for DX than technology:

> Digital transformation is fundamentally not about technology but about strategy. Although it may require upgrading your IT architecture, the more important upgrade is to your strategic thinking. (Rogers 2016, p. 239)

A DX strategy involves prioritizing by picking the problems that matter most for your digital growth agenda, drawing on insights about customer needs. The focus should be on business needs, not on specific technologies (such as blockchain or cloud). All too often, DX is little more than a collection of technology pilot projects that operates in a strategic vacuum. DX efforts can get distracted by chasing the latest flashy new thing. Firms may quickly become overwhelmed with ideas for digital projects, lacking criteria to prioritize which projects to pursue. There is also a danger that DX efforts focus on cost-cutting and optimization of legacy processes, and lacking the strategic guidance into growth (such as launching new products to better meet customer needs).

2.1.1.1 Customers

The first of the five domains refers to customers. In the previous age, characterized by mass production and mass communication, communication with customers was a one-way road, with companies broadcasting their messages and pushing out their products en masse, and little scope for learning from customers.

In the digital area, firms must follow the so-called "customer network model" if they want to stay competitive. In this view, the firm is situated in a network with many different kinds of actors, and interacts with them in various ways. Firms do not merely push out information and products to customers, but interact with various stakeholders through blogs, forums, online comments, social media, YouTube, and so on.

Digital business models have led to a reformulation of the **marketing funnel** (Fig. 2.1). DX has led to new possibilities all along the marketing funnel, in terms of creating awareness (through blogs, activity on internet platforms, etc.); stimulating consideration by customers (e.g. through online reviews); influencing consumer preferences (via social networks and YouTube); encouraging customer action (via several modes: online, in-store, and mobile sales); enhancing loyalty programmes (for example, using social media) and boosting advocacy behaviour from customers (e.g. through product reviews, links, and social media "likes").

Rogers (2016) suggests 5 behaviours that are appropriate for interacting with customers in the digital age, shown below in Table 2.1.

2.1.1.2 Competition

In the previous business age, competition used to take place within defined industry sectors, but now the industry boundaries are more fluid. An example of this could be how Apple solved the conundrum of selling recorded music in the age of digital MP3 files, while the incumbent RIAA (Recording Industry Association of America) seemed unable to thrive in the new technological landscape.

In the digital age, the distinction between allies and rivals is increasingly blurred. This has always been the case to some extent, but DX takes the trend

2.1 Models of Digital Transformation

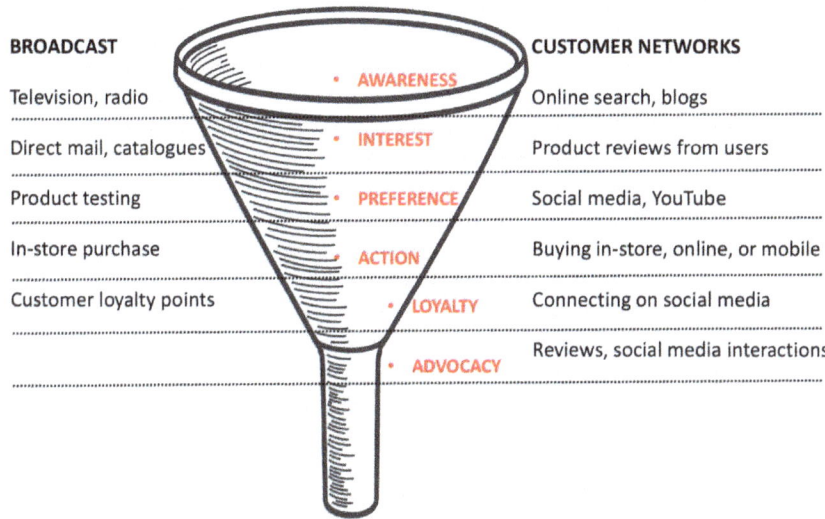

Fig. 2.1 Rethinking the marketing funnel in the digital age. (*Source* Author's elaboration, inspired by Rogers [2016])

Table 2.1 5 behaviours and strategies in the customer network

Firm behaviour	Firm strategy
Access	Firms should always be available, anywhere and everywhere
Engagement	Firms should be seen as a source of valued content
Customization	Firms should adapt their products and services to customer needs
Connect	Firms should connect to customers' conversations
Collaborate	Customers can help to build and shape the business

Source Author's elaboration, inspired by Rogers 2016

further. For example, Microsoft and Apple used to be fierce enemies, but now they are finding common ground, as Microsoft CEO Satya Nadella proudly demonstrated when he held up an iPhone at a trade conference (with Microsoft software installed on it) (Nadella 2017).[1]

Competitors may cooperate in some areas and compete in others. A classic example could be cooperating on basic research in industry-wide R&D consortia, or cooperating on the development of infrastructure, standards and regulations, and then resuming the competitive battle downstream in consumer markets.

---

[1] See also here: https://www.geekwire.com/2015/microsoft-ceo-satya-nadella-keeps-playing-nice-with-bay-area-tech-scene-at-salesforce-conference/ [last accessed 31 July 2025].

In the digital age, key assets are not always held within the firm. One case would be the shifting of data storage over to the cloud, rather than having in-house IT servers. Another case could be how assets can be leveraged through partnerships, without having to own these assets in-house. In the context of winner-takes-all dynamics, that arise because of the increasing returns of network effects, it pays to specialize in a core activity while reaching out to partners in order to access non-core complementary assets.

Another factor affecting the competitive dynamics in the digital age is that many large firms have eschewed the traditional activity of making unique products, to instead providing platforms that are intrinsically cooperative in the sense that inputs from partners play a crucial role in the overall value creation.

2.1.1.3 Data

A primary use case of data analytics is the monitoring of operations, inventory, supply chain, sales, billing, and so on. The monitoring of business processes can lead to efficiency gains and cost reduction. More promising, however, is the use of data analytics to create new value. This can be done by using data to strengthen the main value proposition, for example integrating weather data or maps data into products if this could be useful for customers. Relatedly, customer data could be mined to develop a more thorough understanding of consumer identities, thereby improving the customer experience and leading to higher sales.

Data as an asset should be understood differently in the digital age. The underlying strategic assumptions have changed. Previously, data was expensive to generate, but now it is continuously being generated throughout the organization. Previously, data was managed in operational siloes, now data needs to be accessible throughout the organization. Previously, data was a tool for cost-cutting and optimizing operations, now it is an important tool and valuable intangible asset, for value creation (e.g. launching new products and services to address emerging customer needs). Previously, storing and managing data was the difficult part; but now the challenge is how to convert data into valuable actionable knowledge. In the digital age, data refers to various formats, not just structured data found in traditional data warehouses, but also unstructured data (including text data from online comments and social media, as well as video data).

2.1.1.4 Innovation

Innovation has long been seen as a keystone of competitive advantage. However, DX changes the way that innovation is undertaken. With DX, there is a shift from decisions based on seniority and gut feeling, to decisions based on data analysis and hypothesis testing. Innovation is based on developing minimum viable prototypes (rather than a finished product), and being painfully aware of the need to stay focused on user needs. There is no longer any excuse for launching a product that consumers do not want. In cases where failure occurs, failures are cheap and fast, and failures are learning events.

It helps to distinguish between two types of innovation. The first, "**convergent experimentation**", refers to constant incremental testing in the form of **A/B testing**. Amid growing dissatisfaction with focus groups and committees (which can be expensive, slow, and subject to various self-report biases), there is growing enthusiasm for experiments that involve the actual choices made by people.

A/B testing, in a nutshell, refers to the method of comparing two versions of something to learn which one performs better. A/B testing in the digital age usually refers to online testing of two different formats (e.g. shopping cart buttons, website formatting) to see which one is more popular with customers. Website visitors are presumably unaware that they are participating in an experiment.

The second type of innovation is the case of "**divergent experimentation**". DX changes the way that radical innovation is done, using tools such as rapid brainstorming (perhaps involving generative AI and adversarial learning), rapid prototyping (perhaps using simulations), faster performance feedback, and pivoting the project if it turns out that the overall direction is not aligned with consumer needs.

DX and innovation are also discussed later on in Sect. 8.3.

2.1.1.5 Value

The fifth domain from Rogers' model refers to creating value. The idea is that firms should remain obsessed with their customers, alert and focused on changing customer needs, in order to ensure that their value proposition remains relevant and appealing. DX encourages firms to stay ahead of the competition by looking for the next opportunity for creating customer value (which ultimately leads to higher firm performance).

2.1.2 Schmarzo's BDBMMI

Schmarzo (2016) puts forward a 5-stage model called the Big Data Business Model Maturity Index (BDBMMI), shown below in Fig. 2.2. The BDBMMI model seeks to measure the maturity of firms regarding how data and analytics are integrated into their business models. The BDBMMI model also aims to map and benchmark the positioning of enterprises with regards to industry best-practice, and also provide a roadmap for further leveraging data and analytics. The BDBMMI's first 3 stages focus on internal process optimization, drawing on data from operations, products, and customers. Then, the BDBMMI's last 2 stages (Data Monetization, and Business Metamorphosis) are oriented towards external market opportunities, applying the insights from analysis of customer, product and market data.

2.1.2.1 Business Monitoring

This stage refers to the starting point of the DX journey. Firms collect basic analytics in the form of standardized, retrospective reports, generated from the firm's data warehouse, to monitor the organization's performance and to signal past performance. Business Intelligence is backward-looking (e.g. financial results from

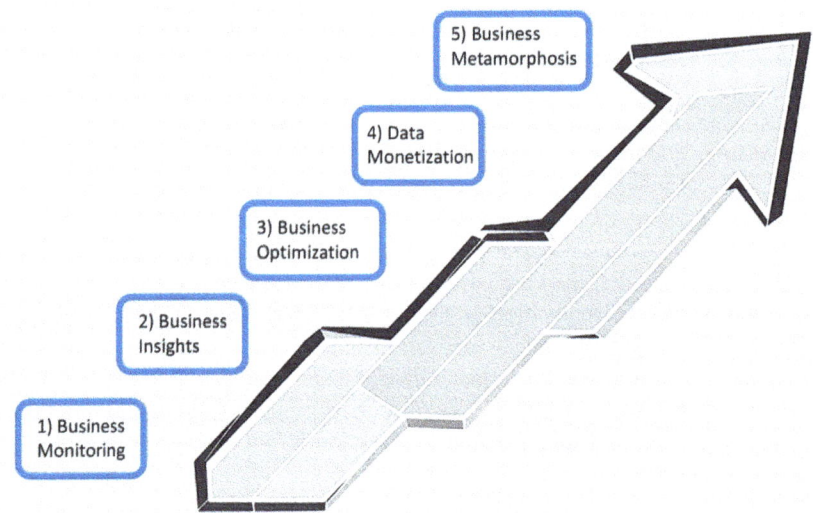

Fig. 2.2 The Big Data Business Model Maturity Index (BDBMMI). (*Source* Author's elaboration, inspired by Schmarzo [2016])

the previous completed period) and the situation is compared to driving your car while looking in the rear-view mirror.

2.1.2.2 Business Insights

The second stage corresponds to firms that have taken some first steps away from the backward-looking traditional practices (such as standardized reports on past performance), towards gleaning new insights from their data assets.

The goal is to move from monitoring the past, to making **predictions** about the future (or at least having more accurate predictions of the present using real-time data, sometimes referred to as "nowcasting"). Better still, if a firm has some useful predictions about the future, it can seek **prescriptions** which are recommendations for appropriate behaviour, assuming that the predictions are reasonably accurate.

Firms can also start accumulating the data that they might need, preferably real-time data, in the form of new sources of internal data (such as readings from machines and smart devices, emails, technicians' notes) as well as external data (such as weather data, external events, and social media posts).

This second stage of the BDBMMI model corresponds to "crossing the analytics chasm" (Schmarzo 2020), which corresponds to a significant shift in perspective. The firm moves from monitoring the past, towards seeking predictions and prescriptive analytics about the near future. With regards to data analytics, a deeper appreciation of the value of data leads to storing a wider variety of data formats (rather than specific formats being stored in a highly-regulated data warehouse), storing not only aggregate data but also detailed disaggregated data, and opening up access to data for real-time processing and data analytics by a broader range of possible users.

2.1.2.3 Business Optimization

In the third stage of the BDBMMI, the firm seeks to optimize based on the predictive and prescriptive data analytics. Progress is made in moving beyond backwards-facing reports, towards forward-looking insights to employees and managers, to respond to questions such as "Tell me what I need to do." As such, actionable insights are delivered to employees, partners, and other stakeholders to help optimize their decisions. Where possible, predictive and prescriptive analytics are integrated into routines and embedded into processes, such as dashboards to support nowcasting and decision-making.

An example could be Amazon's "Customers who bought this item also bought…" feature, which uses data analytics insights to deliver relevant actionable suggestions for customers at the time of purchase. Another example could be the delivery of analytics-based recommendations to retail store managers to optimize price reductions based on purchase patterns, weather, local events, inventory, holidays, sales promotions from nearby rivals, and so on.

2.1.2.4 Data Monetization

Stage 4 in the BDBMMI corresponds to a stage of **data monetization**, where the increasing sophistication of predictive and prescriptive analytics allows firms to not only improve the efficiency of existing operations, but enable the creation of new value via new product offerings. Data-driven insights can be leveraged to create new sources of revenue. Analytics can be integrated into a new generation of "smart" products and services. Insights into customer behaviour could stimulate entry into new regions or product categories. An example could be how the online market platform eBay improved its service by providing recommendations to the platform's small business merchants.

In some cases, the data-driven insights, or the data itself, can be sold to others (under the constraints, of course, of complying with data confidentiality regulations).

2.1.2.5 Business Metamorphosis

The fifth and final stage in Schmarzo's BDBMMI model is called "Business Metamorphosis". The firm has accumulated rich data assets, it has sophisticated predictive and prescriptive data analytics and data science capabilities to make sense of the ever-changing environment, and has integrated these analytics insights into products and services. Stage 5 corresponds to the case of the transition from a manufacturing firm to a service provider.

An example of the shift from manufacturer to service provider comes from Phillips, which was founded as a light-bulb manufacturer in 1891, but which has transitioned to seeing itself as a service provider in the healthcare sector (Leinwand and Mani 2022). Phillips now focuses on the customer experience, setting up compensation packages with regards to successful patient outcomes (rather than volume of procedures performed), and regardless of whether the treatment uses its own machines, or uses machines made by its partners. This helps Phillips to achieve its stated goal, "to make the world healthier and more sustainable, with the goal of improving the lives of 2.5 billion people a year by 2030."

Another example of a firm in the 5th stage of Schmarzo's BDBMMI, having completed the transition from manufacturer to analytics-based service provider, is the story of the Japanese firm Komatsu, maker of construction machinery (Leinwand and Mani 2022). Komatsu traditionally sold advanced machines as a "product out" strategy, measuring performance in terms of numbers of units sold. A turning-point came in 2000, when setting up a rental business took them closer to customers and yielded new insights. Komatsu observed that the new machines were not leading to the expected productivity gains, because bottlenecks at the construction site prevented these machines from being used in the best way. In technical terms, Komatsu's machine could remove and dump 50% more dirt, but in practice, these gains did not materialize, because construction companies found it difficult to forecast and schedule the dump trucks for the removal of dirt from the construction site. The bottleneck was a problem of information and coordination, and data analytics could play an important role. This contrasts with the traditional solution of focusing on the manufacture of higher-performance machines, which could only have limited gains because this could only affect a few processes on the worksite. As such, Komatsu launched its "Landlog" platform: which included high-resolution drone surveying, mapping geographical data in 3 dimensions, a construction work planning tool, and digital twins for building sites. With the landlog platform, drones take 20 min for surveying (instead of 3 days); and drones feed information into automated bulldozers. Komatsu therefore achieved remarkable productivity growth by focusing not just on increasing the technical performance of its machines, but understanding the broader context, and setting up an ecosystem of suppliers and stakeholders to better connect and coordinate among the people and companies involved in the construction work.

Firms in the fifth stage of the BDBMMI must also have been successful at creating a culture that encourages continuous exploration, as well as the creation, collaboration, reuse, and refinement of an organization's digital assets. Culture is important at all stages of digital maturity, of course, but it is especially important in Stage 5.

To summarize, Schmarzo's 5-stage BDBMMI model provides a useful framework for understanding digital transformation, focusing on themes such as analytics, data monetization, and the transition towards selling services rather than physical goods. We now turn to the 4-stage model of Iansiti and Lakhani (2020) which focuses more on the role of IT architecture, data hubs, and the role of AI.

2.1 Models of Digital Transformation

2.1.3 Iansiti and Lakhani (2020)

The four-stage model of Iansiti and Lakhani (2020, pp. 118–120) starts with the "Siloed data" stage, before moving to the "Pilot" stage, with the firm then becoming a "Data hub", and eventually transforming into an "AI factory." The final stage of AI factory is an impressive creature: an automated entity which is tended by employees, but the employees are external to the actual production process:

> Ultimately, in a digital operating model, the employees do not deliver the product or service; instead, they design and oversee a software-automated, algorithm-driven digital "organization" that actually delivers the goods.

Figure 2.3 shows these four stages.

2.1.3.1 Siloed Data

The starting point of the 4-stage model is a large firm with siloed data, for example a traditional multidivisional business. In this firm, the data are highly fragmented, is located in various siloes, with problems with as inaccuracies in the data, incompatible data structures, and a lack of common identifiers that would enable linking data across siloes (Iansiti and Lakhani 2020, p. 62).

An example could be a hotel chain, with separate customer databases for each country. Customers could become annoyed if they have to repeatedly fill in their personal details, explain again their preferences, and if they find out that their customer loyalty points are not transferable from one country to the next.

While the organization of multidivisional firms into siloes may have been an efficient arrangement in the twentieth century, things have changed in the digital age.

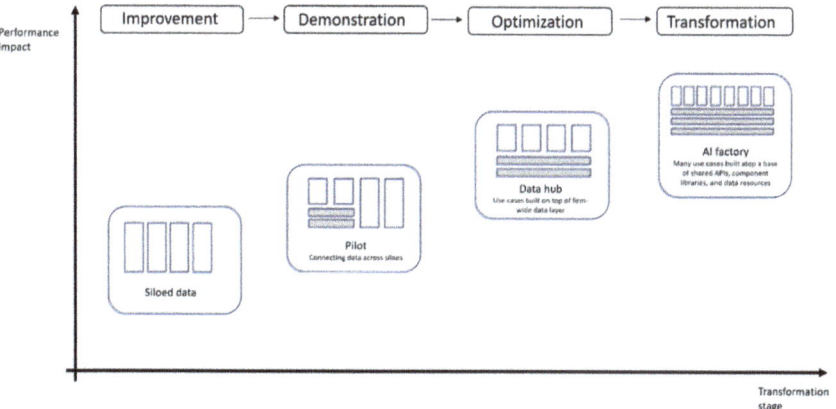

Fig. 2.3 Four stages of digital operating model transformation. (*Source* Author's elaboration, inspired by Iansiti and Lakhani [2020, p. 119])

2.1.3.2 Pilot

Firms move into the Pilot stage (stage 2) without much resistance or opposition, because the value of DX in general is well-known, and the commitment to the new organizational structure is still small and tentative. Efforts are made to connect data across siloes, but these changes do not disrupt ongoing operations in other parts of the organization. Stage 2 seeks to demonstrate the potential of DX, while full-scale application comes later. Firms entering Stage 2 hope to soon keep going in the transition to more sophisticated stages.

2.1.3.3 Data Hub

Broad-based reorganization takes place in the Data Hub stage. The many data siloes are connected to create a single data resource that spans the entire organization. Considerable investment is needed, and at this stage resistance to change is also likely to increase.

> One of our biggest surprises in transformation efforts (maybe obvious in retrospect) is the frequent resistance of the CIO and of the IT organization. (Iansiti and Lakhani 2020, p. 112)

The reasons for resistance from the CIO (Chief Information Officer) and IT department could be because IT departments were originally designed for a different purpose (i.e. back-office jobs, not innovation and transformation). Traditional IT skill sets rarely include data science, advanced analytics, machine learning, and AI. As data analytics moves from the production of standardized backwards-facing reports, to the domain of advanced data science techniques to produce forward-looking analytics on an exploratory basis, this can lead to changes in the organizational culture regarding the role of data.

2.1.3.4 AI Factory

The "AI factory" stage refers to the concept of a factory without people: or at least, a factory where the employees tend to the production processes from outside while the main production work is performed by machines. An example could be Ant Financial Services Group, the Alibaba spinout that uses AI in combination with Alipay data. Ant Financial is built on a digital core, and the value generated by its processes (such as consumer lending, wealth management, and credit rating services) is delivered by AI (not human employees). Human workers do not appear in its "critical path" of operating activities (Iansiti and Lakhani 2020).

The AI factory is organized around small teams. The emphasis on small teams does not imply division into siloes, however: the data is centralized, and cross-team communication and collaboration are carefully designed and built into the teams' activities.

Firms at the "AI factory" stage cannot rest on their laurels. Innovation does not stop: instead these firms are now well-positioned for an ongoing journey of continuous capability-building and innovation.

2.1.4 Comparing these Models

These three models, taken together, provide different yet complementary insights into the process of digital transformation.

Rogers (2016) was an influential early model that combines five important domains into an integrative and coherent whole: Customers, Competition, Data, Innovation, and Value. The ideas arranged into these 5 domains have been echoed in many DX publications by other scholars. While Rogers (2016) does not arrange these 5 domains in a temporal ordering in terms of stages, nevertheless this is attempted in Schmarzo (2016) and Iansiti and Lakhani (2020).

Still other models of the stages of the DX process have been put forward. Bonnet (2022)'s short article contains a three-stage model (modernization, enterprise-wide transformation, and new business creation). Kane et al. (2019, pp. 220–221) suggest a 4-stage model of digital maturity (exploring digital efforts, doing digital initiatives, becoming digitally mature, and being a digital organization). Davenport and Harris (2017, p. 61) present a 5-stage model of maturity regarding digital analytics (analytically impaired; localized analytics; analytical aspirations; analytical companies, and analytical competitors). These models are interesting additions to the literature, although they overlap to some extent with the main ideas discussed in the chapter, and are not explored in depth here.

Models of the stages of DX are useful approximations, although they should not be taken too literally in empirical investigations because of difficulties in precisely defining the steps, and the possibility that some firms could skip over steps, or face the same steps many times. Furthermore, digital transformation may look different when comparing one firm to another, to the extent that managers should not merely try to copy what worked elsewhere (or even what worked elsewhere inside the same firm), but that they should be attentive to the performance of digital initiatives by comparing with carefully-designed metrics.[2]

Schmarzo's model suggests that firms in advanced stages could engage in data monetization, and specifically to sell data (or data analytics) to third parties. This eventuality does not feature directly in the model of Iansiti and Lakhani (2020), perhaps because of growing concerns about data privacy and confidentiality. Schmarzo's model also emphasizes business-as-a-service as the ultimate state of DX, while Iansiti and Lakhani present the ultimate state in terms of an autonomous AI-operated factory. Iansiti and Lakhani focus more prominently on AI, perhaps because AI has advanced rapidly in industrial applications over the intervening period (from Rogers and Schmarzo in 2016, to Iansiti and Lakhani in 2020).

The threat of digital to traditional firms is real and large. However, the appropriate response need not require jumping in and betting the company on a dramatic strategic pivot. Instead, an incremental approach is recommended (McGrath and

---

[2] Of course, this does not mean that standardized metrics used by rivals, or used by top-performing superstars, should be blindly copied and applied in a new context.

McManus 2020). A common theme in Schmarzo (2016) and Iansiti and Lakhani (2020) is that the DX journey should start by focusing on a relatively short-run business initiative (or "pilot"), that can yield a quick win after about 9–12 months, as a way of demonstrating the potential of DX, and inspiring employees and managers to continue further down the DX path. In a similar way, Rogers (2023) emphasizes the importance of leading the firm through DX with a shared vision, and also starting the DX journey by prioritizing a subset of business initiatives that matter the most.

Schmarzo (2016) also recommends the "Data Lake" architecture, which is certainly an improvement over a siloed data architecture, although Data Lakes might not always be the best solution in all cases (Lamarre et al. 2023). This is discussed further in Sect. 3.2.

Finally, the existence of a fixed number of stages in these models does not imply that "the work is done" after reaching the final stage. Instead, DX is an endless journey. Rogers (2023, p. 5) defines Digital Transformation this way: "Transforming an established business to thrive in a world of constant digital change."

Further Reading

Further details on the three models of digital transformation discussed in this chapter can be found in Rogers (2016, 2023), Schmarzo (2016, 2020) and Iansiti and Lakhani (2020).

References

Bonnet, D. (2022). 3 stages of a successful digital transformation. Harvard Business Review. 20th Sept 2022. https://hbr.org/2022/09/3-stages-of-a-successful-digital-transformation

Davenport, T., & Harris, J., (2017). Competing on Analytics: The New Science of Winning. Harvard Business Review Press, Boston, MA, USA.

Iansiti, M., & Lakhani, K. R. (2020). Competing in the age of AI: Strategy and leadership when algorithms and networks run the world. Harvard Business Press.

Kane, G. C., Phillips, N., Copulsky, J. R., & Andrus, G. R. (2019). The Technology Fallacy: How people are the real thing to digital transformation. MIT Press.

Lamarre, E., Smaje, K., & Zemmel, R. (2023). Rewired: The McKinsey guide to outcompeting in the age of digital and AI. John Wiley & Sons.

Leinwand, P., & Mani, M. M. (2022). Beyond Digital: How Great Leaders Transform Their Organizations and Shape the Future. Harvard Business Review Press: Cambridge, MA.

McGrath, R., & McManus, R. (2020). Discovery-driven Digital Transformation. Harvard Business Review, 98(3), 124–133.

Nadella, S., (2017). Hit Refresh: The Quest to Rediscover Microsoft's Soul and Imagine a Better Future for Everyone. HarperCollins publishers, New York, NY.

Rogers, D. L. (2016). The digital transformation playbook: Rethink your business for the digital age. Columbia University Press.

Rogers, D. L. (2023). The Digital Transformation Roadmap: Rebuild Your Organization for Continuous Change. Columbia University Press.

References

Schmarzo, B. (2016). Big Data MBA: Driving business strategies with data science. John Wiley & Sons.

Schmarzo, B. (2020). The Economics of Data, Analytics, and Digital Transformation: The theorems, laws, and empowerments to guide your organization's digital transformation. Packt Publishing Ltd.

Big Data Technologies and Architecture 3

3.1 A Historical Perspective on Appropriate Organizational Structure

The history of economic development has seen the rise and fall of various ways of organizing economic activity (Dosi 2023, Sect. 4.3.2).

In the pre-industrial age, the production of textiles relied on manual labour by small-scale artisans, in the context of a rural **cottage industry**, where workers stayed at home in their cottages had autonomy and flexibility in choosing their working hours, and would get paid by bringing their produce to be sold at markets.

The factory system, which emerged in the eighteenth and nineteenth centuries, changed life in many ways: the shift towards urbanization, the multiplication of schools and churches, and change in lifestyles (such as the new time-discipline imposed upon factory workers, who could no longer choose when they would wake up; as well as a loss of autonomy regarding work processes and methods). Key economic advantages of the factory system came from standardization of business processes, specialization within the factory system, and the setting up of production lines.

Much later, in the twentieth century, came the rise of the large **multidivisional corporation**. A defining feature of the modern corporation was the multidivisional ("M-form") structure, which refers to there being many divisions (each focusing on a particular product or geographical unit), that operate under the strategic guidance of a corporate head office. Production efficiency benefitted from having the various divisions exercising operational control over their respective areas. The company was managed via functional divisions (finance, sales, R&D, and so on).

Supplementary Information The online version contains supplementary material available at https://doi.org/10.1007/978-981-95-2433-4_3.

© The Author(s), under exclusive license to Springer Nature Singapore Pte Ltd. 2025
A. Coad, *Data Science MBA*, Springer Texts in Business and Economics,
https://doi.org/10.1007/978-981-95-2433-4_3

Technical details of the production processes were overseen by middle managers in the various divisions. The economic advantages of the modern multidivisional corporation include economies of scale, while allowing for the benefits of specialization as co-workers operate in close proximity, sharing knowledge largely within their own divisions.

A catchphrase of the digital era is that managers are exhorted to "break down siloes". If siloes are so bad, how did they get there in the first place? **Siloes** actually used to be best-practice, in the age of the M-form organization. In the context of specialization of human resources, and also in recognition of the limits to the distances over which humans can effectively communicate, it used to make sense to organize employees into specialized divisions where they could process information and engage in teamwork and business process routines by simply swivelling their office chairs and interacting with colleagues. When sharing information was costly, it made sense to only share information within dedicated functional siloes.

However, siloes are not best-practice in the digital era! It is now relatively costless and trivial to replicate data and send it around the world. Physical proximity is not a binding constraint for machines interacting with machines to process data. Instead, there are gains to sharing data across the entire boundaries of the organization, even for global corporations. That said, productive activity in a modern firm seems to be best organized by small agile teams, modular in organization yet connected with various internal and external partners. Small agile teams operate in a context of continuous innovation while drawing on company-wide data assets, such as a data lake (Iansiti and Lakhani 2020).

3.2 A Single Data Lake

The previous era belonged to large multidivisional firms, arranged into separate divisions (according to product lines or regions) that were operated as independent profit centres under the strategic direction of the headquarters. Multidivisional firms often organized their activity (including their rudimentary data assets) into siloes corresponding to the organizational divisions. In such organization, it may be the case that "data is power", and that managers and employees are reluctant to share their data without negotiations and asking for favours. In the digital age, however, an organizational architecture characterized by siloes becomes problematic, and there is a need to redesign a firm's organizational structure. Storing data in siloes hinders the flow of information across the organization, leading to a failure to spot connections across departments, and preventing firms from having a 360 degree view of the customer. Instead of confining databases into siloes, there should be a single **Data Lake** that serves as a single unified infrastructure resource all across the firm. The "AI factory" model of (Iansiti and Lakhani 2020) in Sect. 2.1.3 highlighted the need to connect data across siloes to have a centralized data repository.

The term "**Data Lake**" was introduced by James Dixon, CTO of Pentaho (Dixon 2010) as a solution that handles raw data from one source and supports diverse user requirements (Hai et al. 2023). A data lake stands in sharp contrast to the alternative concepts of **data warehouses** or **data marts** for which the structure and usage of the data must be predefined and fixed, and rigorous data extraction, transformation, and cleaning are necessary before entering data.

Data warehouses were popular in situations such as where management wanted standardized and highly-structured backward-looking reports showing summary statistics and aggregates (such as sales per region) for the most recent previous periods. However, a drawback of data warehouses comes from their regulated way of receiving new data. Data must be of high quality (in the sense of being free from error), in standardized format, and structured in a way designed for creating standardized reports. This leads to expensive pre-processing before data is entered into a data warehouse, and as a result most organizations would limit how much data they would store in their data warehouse (Schmarzo 2016). In contrast, data lakes store data in the original format, rather than entering data into a standardized format, which makes it easier to ingest and store data, and allows organizations to store a large variety of different types of data (numbers, text, images, video, etc.). Data lakes are therefore more suitable for storing a variety of structured, semi-structured, and unstructured data coming from a variety of sources such as business transactions, sensors, or mobile/cloud-based applications. Data lakes can also be worked upon by users who add value to the original data by cleaning, removing errors, labelling variables, creating links, adding domain-specific knowledge, and providing additional descriptive information via metadata. As such, there is a dynamic interaction between the data lake and users, in order to continuously improve the quality and the value of the data. Data lakes can also be built using tools such as commodity hardware and Open Source database management software (such as Hadoop), as well as free online training for big data architectures, which allows significant cost reductions for storing data (Schmarzo 2016). While data warehouses were useful in situations where managers only wanted regular standardized reports on recent performance indicators, the data lake is more useful in situations where a variety of databases can be analyzed on-demand by a data science team that focuses on exploring hypotheses about current and future evolutions of their customers and markets. As such, a data lake is a shared storage platform that natively supports both traditional and next generation workloads (Schmarzo 2016).

A data lake is therefore a massive collection of datasets that may vary in their formats; may be hosted in different storage systems; may not necessarily be accompanied by any useful metadata (or may use different formats to describe their metadata); and that may change autonomously over time (Nargesian et al. 2019). A data lake is a useful solution that helps to reduce costs (because it is inefficient to store multiple versions of the same data) and also avoids the problem of having multiple (conflicting) versions of datasets. Data lakes help organizations to have a **"Single Source of the Truth"** (SSOT, DalleMule and Davenport 2017), thereby avoiding the nightmare scenario where managers are unable to communicate properly because they are arguing about different numbers for key indicators that have different definitions.

A maxim for how to increase workplace productivity is that "there is no reason ever to have the same thought twice, unless you like having that thought" (Allen 2015). A similar idea holds for organizations: there is no need for any employee to clean and manipulate data that has already been done by someone else. Data lakes help improve productivity by avoiding such repetitions.

Data quality is an important area for organizational performance (Nagle et al. 2020). That said, data lakes pose a daunting challenge for database management, because of the huge volumes, large variety, and rapid velocity of incoming data. As such, a potential problem for data lakes (compared to data warehouses that ingest highly-structured data) is that ingesting disparate data might easily turn the data lake into an unusable **data swamp**, unless there are procedures for effective metadata management and data governance. In particular, after ingestion, the semantics and data quality of the raw data are unknown, and the origin (provenance) of individual datasets and its possible connection to other datasets could be missing. According to the **FAIR principles** (Hai et al. 2023, p. 8), the content of a data lake should be Findable, Accessible, Interoperable, and Reusable.

Finally, a data lake is not merely a storage repository, but it is a strategic asset that is easily scalable, and therefore plays a crucial role in positioning firms for growth. Besides data lakes, another solution for firms engaging in digital transformation is cloud computing, which offers further benefits in the area of cost savings, flexibility, and scalability.

3.3 Cloud

A powerful case for **cloud computing** is made by the following statistics. How much of a computer's resources does the average person use during peak usage times on an average workday? The average user uses about 10% of the processor, less than 60% of memory, and 20% of network bandwidth. Then multiply this by the thousands of employees in a large firm, and the resulting efficiency gains from moving to cloud are striking (Ruparelia 2023).

A major advantage of cloud computing is that it allows firms to better manage their computer resources, to better align the actual use of resources with what they are paying for. Cloud therefore allows cost savings from a more efficient utilization of resources. Cloud can benefit users in terms of flexibility (if the needs for computational resources fluctuate over time), and scalability and agility (if a rapidly-growing firm needs to suddenly double in size). Cloud can also lead to improved security, enhanced collaboration opportunities, and also access to advanced technologies (for example, services sold in the cloud). Furthermore, having data in the cloud enables the creation of AI applications that integrate, analyze, and derive insights from heterogeneous data sources. A major drawback, however, is that the process of migrating systems to the cloud can be messy and can take several years.

SMEs will probably buy their cloud services from an external service provider such as AWS (Amazon Web Services), although large firms might prefer the option of setting up their own cloud infrastructure for their own use.

Cloud computing therefore provides a useful solution for managing and scaling digital resources. The next Section discusses how APIs and microservices offer additional benefits for digital transformation, such as modularity, flexibility, and innovation.

3.4 APIs and Microservices

APIs (Application Programming Interfaces) enable a change in architecture that allows firms to become more flexible and capable of continuous upgrading and innovation, as well as encouraging the sharing of data with external stakeholders. A vivid illustration of the advantages of APIs (sometimes referred to as a **microservices** architecture) comes from Amazon.

Around the year 2002, Amazon was facing problems brought on by its own rapid growth. The monolithic software infrastructure that supported its processes was under strain because of the number of different products and businesses that were held together in an unreliable way. As a result, CEO Jeff Bezos sent out something resembling the following memo, which is now a well-known event in the history of DX (Box 3.1). This memo outlined a new architecture based around APIs and microservices, that turned out to be crucial in enabling modularity, flexibility, and ultimately in enabling innovation.

> **Box 3.1: Bezos's API Mandate**
>
> All teams will henceforth expose their data and functionality through service interfaces. Teams must communicate with each other through these interfaces.
>
> There will be no other form of inter-process communication allowed: no direct linking, no direct reads of another team's data store, no shared-memory model, no back-doors whatsoever. The only communication allowed is via service interface calls over the network.
>
> It doesn't matter what technology they use.
>
> All service interfaces, without exception, must be designed from the ground up to be externalizable. That is to say, the team must plan and design to be able to expose the interface to developers in the outside world. No exceptions.
>
> Anyone who doesn't do this will be fired. Thank you; have a nice day!

Source https://apievangelist.com/2012/01/12/the-secret-to-amazons-success-internal-apis/ [last accessed 27 Sept 2024]; see also Iansiti and Lakhani (2020, p. 79).

The Bezos memo announced a dramatic shift in architecture, the seriousness of which was underlined by the last line, that threatens to fire anyone who does not comply. Rather than allowing teams to share data and resources with each other within the firms' boundaries, they had to interact with each other through internet-based services operating via APIs. As a result, teams had to decouple, define which resources they had, and make these resources available through an API. This led to a whole new set of challenges, such as support for a team's API interface, security issues, monitoring, testing and debugging, and so on. Amazon overcame these challenges, and the shift to APIs and a microservices architecture was pivotal in the later success of Amazon (such as the launch of the hugely successful Amazon Web Services).

Let us first consider how APIs can provide advantages, even if only used internally, within a firm's boundaries. APIs enable modularity within the firm, spurring a shift from a monolithic IT architecture to a modular architecture (see Table 3.1). The modular architecture allows firms to implement changes in one partitioned module, without having to commit changes to a single monolithic block of code. As such, a modular architecture allows for the development, verification, and implementation of new software code (i.e. "DevSecOps") to be done independently, speeding up responsiveness and innovation, while reducing the risk of system failure.

> Amazon can deploy software thousands of times a day ... By contrast, many large companies are limited by inflexible architecture: no matter how fast they can code programs, they can deploy software only a few times a week or a month. (Rigby 2020, p. 158)

There may be hundreds or thousands of microservices within a single firm. As such, there will be a need for capable managers who can see across the whole system with a global view of how the APIs are interconnected.

Another advantage of APIs is that they open up new possibilities for firms to interact with external partners. APIs allow data to be shared in a secure way (through an API gateway that the firm controls). APIs are simple in the sense that it does not require large efforts in programming an interactive application, or manually checking which data is being accessed by which external partner. APIs therefore open up new business opportunities in areas such as using data across a variety of systems and use cases, monetizing data assets in a secure way, and allowing external partners to check up on your inventory. Airbus provides an interesting case regarding how APIs open up new possibilities for sharing data, even data that is sensitive (Wang and McLarty 2021). Airbus recognized an issue in the aviation industry where critical flight and operational data was locked away in siloes. From its position as a supplier to most airlines and thus having a strong

Table 3.1 How microservices enable the shift from Monolithic to Modular IT

Monolithic IT	Modular IT
One integrated program (like when building an MVP)	Microservices as an architectural approach to building applications that partition apps into services that can be built, tested, and updated independently
Traditionally, any update required a full version update that impacted the work of every associated team	Separate services that are loosely coupled through automated interfaces: APIs
To change one part, you must test on the whole system or risk bringing it all down	Hundreds or thousands of microservices within a single firm
Bad for speed, quality and flexibility	Continuous development & improvement across services
	Need for someone who can see across the whole system

Source Author's elaboration, based on Vaz (2021, p. 116) and Rogers (2023, p. 227)

relationship across the ecosystem, Airbus led the way in convincing airlines to cooperate with the launch of an API-enabled data platform called Skywise to help the airlines reduce maintenance issues and prevent technical delays.

The initial shift from monolithic IT to modular IT was challenging for many firms. The challenge these days, interestingly, that some firms are so comfortable with APIs that they are using too many. Lamarre et al. (2023, p. 176) observe that "companies tend to create too many APIs" and that the recommendation is often to "[m]inimize the number of APIs and optimize their use."

3.5 Small "Agile" Teams

Agile is a management philosophy that refers to the advantages of organizing complex, uncertain projects in self-managing multidisciplinary teams. The concept of "agile" started in the manufacturing sector, but is now often associated with digital-native firms such as Spotify and Amazon (Rigby et al. 2020).

An early perspective from Takeuchi and Nonaka proposes a "rugby" approach: "a team tries to go the whole distance as a unit, passing the ball back and forth." The team works together, interacting intensely and bringing their functional expertise to help the team to quickly maneuver through an uncertain field to bring the project to its goal. Team members can be hand-picked from several different departments. For example, one team member might come from Marketing, another from Regulation and Compliance, another from Supply Chain, another from Manufacturing, another with a background in DevOps (Software Development and IT Operations), alongside a team member with experience coaching agile teams. This team can satisfy requirements from various perspectives (e.g. manufacturing, marketing), avoiding the usual stoppage times that affect communication between departments, to develop a credible new product as a solution to a specific problem (such as a customer pain point, or a specific aspect of the customer experience) with unmatched speed. The agile team develops a **Minimum Viable Prototype (MVP)** which is tested on consumers in order to quickly iterate towards a legitimate new product. Agile teams work with prioritized do-lists ("backlogs") and engage in time-limited work cycles of around 2–4 weeks ("sprints").

Agile can be expected to not only increase team productivity, but also reduce time spent in meetings, reduce the resources spent on planning and documentation, improve the efficiency of resource utilization (e.g. if under-performing projects are killed early), increase the satisfaction of employees as well as customers, and reduce risks such as the danger of releasing an irrelevant product to market (Sutherland 2014; Rigby et al. 2020).

Agile can also lead to problems, if not done right (Rigby et al. 2020). For example, there may be resentment that star employees are being poached from their respective departments, with traditional budgeting procedures and management practices being disregarded, thereby potentially leading to confusion and chaos. Nevertheless, there is ample evidence that agile, done right, can reliably lead to superior performance.

While small agile teams provide a useful solution for organizing and managing digital projects, the next Section discusses how the analytics sandbox environment offers additional benefits for digital transformation, such as experimentation, collaboration, and innovation.

3.6 Analytics Sandbox Environment

The data lake serves as a useful piece of infrastructure for the whole organization. But what happens if a team tries to improve the data, through activities such as cleaning and improving the core data, generating derived data or developing new prediction models? Such outputs should be made available to the whole organization (to avoid repetitions, and to facilitate the reuse of repurposable components of software code, etc.), but during the time that they are being developed this should be done outside of the Data lake environment. The solution is to have an **analytics sandbox** environment that is set up alongside the data lake (Schmarzo 2016).[1] Such an environment serves as a platform for sharing tools, models, data products, and repurposable software code components across the organization (Ross et al. 2019).

In contrast to the regular production of standardized reports and data dashboards to help management monitor operations, the analytics sandbox is an ad-hoc, fail-fast environment where data science teams can explore hunches and hypotheses to seek out valuable new insights and business opportunities. Data science teams working in an analytics environment can quickly ingest and analyze new datasets,

---

[1] The analytics sandbox is similar to the concept of "digital platform" in Ross et al. (2019).

3.6 Analytics Sandbox Environment

Table 3.2 Comparing the data lake with the analytics sandbox environment

	Data Lake	Analytics Sandbox
Routinization	Creation and delivery of scheduled standardized reports	Exploring ad hoc hypotheses that may lead to new revenue opportunities
Concepts of "Truth"	Single Source of the Truth (SSOT)	There may be multiple alternatives (e.g. exploring alternative future scenarios, developing various predictive models)
Technology requirements	Predictable IT system load and software. Stable, scalable, and secure systems	System load and software depends on the data scientist team, as they explore new hypotheses, new data sources, new techniques, etc. Repositories of reusable API-enabled components to be shared among data science teams throughout the organization
Governance	Heavily governed, to ensure the historical data is accurate	Loosely governed, not yet sure if the data science outputs will be valuable

and apply advanced software tools (such as R and Python) in the hope that some new valuable and unexpected insights may arise. Table 3.2 contrasts the data lake with the analytics sandbox environment.

Data science teams working in the analytics sandbox environment may enrich the core data in various ways, such as cleaning the data, resolving errors, labelling variables, creating links between databases, adding domain-specific knowledge, and providing additional descriptive information for databases via metadata. These valuable activities should eventually result in the data lake being updated. Furthermore, data science teams working in the analytics sandbox environment may develop sophisticated tools (such as software components, predictive models and algorithms) that could potentially be useful in other parts of the organization. For example, Toyota developed a system of connecting vehicles to the cloud, which resulted in a capability of tracking cars. This car-tracking capability was initially developed for the purposes of car sharing, but the car-tracking capability could be rapidly repurposed for tracking buses travelling between cities (Ross et al. 2019, p 62). There is a need for a platform that connects teams across the organization, such that these kinds of potentially-reusable tools (such as modular software components) can be shared, to avoid repetitions and improve efficiencies. In this case, of course, new capabilities should not be developed as monolithic pieces

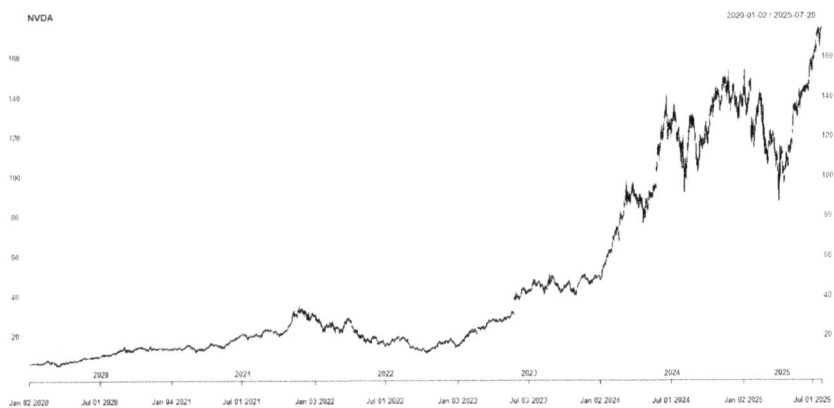

Fig. 3.1 Stock price dynamics for Nvidia. (*Source* Author's calculations using the quantmod package, see R code file for details)

of code (setting up the **technical debt**[2] of tomorrow), but as API-enabled modular components that have been designed to facilitate being repurposed across applications.

3.7 R Example: Data from an API

Figure 3.1 shows the stock price dynamics for Nvidia, up until the time of writing. The quantmod package can retrieve recent data from online sources using APIs. The quantmod procedure essentially uses the Yahoo Finance API (or other financial data APIs) under the hood to fetch the data. As such, the user typically does not need to handle API requests themselves. The quantmod package documentation and related resources often describe the interaction with data sources like Yahoo Finance in terms of data retrieval methods rather than explicitly using the term "API." However, the underlying mechanism involves using APIs provided by these financial data sources.

Another example of the use of APIs in R is the package IBrokers,[3] which provides native R access to Interactive Brokers Trader Workstation API. At time of writing, IBrokers is only able pull data from the Interactive Brokers servers via the TWS, although future additions are expected to include more API access,

---

[2] Technical debt refers to future costs that the business incurs because of suboptimal technology (Rogers 2023). Examples can include poor HTML code that slows down a webpage, to deficient infrastructure for networking or cybersecurity. There can be many causes of technical debt, such as deferred maintenance of old systems, changes in technology standards, and poor initial design (which may originally have been made with a "move fast and fix it later" mindset; Rogers 2023). Lamarre et al. (2023, p. 198) identify 11 different types of technical debt.

[3] https://cran.r-project.org/web/packages/IBrokers/index.html [last accessed 26 July 2025].

including live order handling. Real-time charting (via the `quantmod` package) may also be incorporated into future releases.

Yet another example of how R uses APIs is the "`googleLanguageR`" package (Edmondson, 2020), which calls Google Cloud machine learning APIs for tasks such as the detection and translation of text, sentiment analysis, and the conversion of sound files to text (and vice versa). Text analysis will be discussed in Chap. 11.

Further Reading

Lamarre et al. (2023)'s book focuses on the "how to" questions of digital transformation, giving rich insights regarding big data technologies and architecture.

References

Allen, D. (2015). Getting things done: The art of stress-free productivity. Penguin. Nagle, T., Redman, T., & Sammon, D. (2020). Assessing data quality: A managerial call to action. Business Horizons, 63(3), 325–337.
DalleMule, L., & Davenport, T. H. (2017). What's your data strategy. Harvard Business Review, 95(3), 112–121.
Dixon, J. (2010). Pentaho, Hadoop, and Data Lakes—James Dixon's Blog. https://jamesdixon.wordpress.com/2010/10/14/pentaho-hadoop-and-data-lakes/
Dosi, G. (2023). The Foundations of Complex Evolving Economies: Part One: Innovation, Organization, and Industrial Dynamics. Oxford University Press.
Edmondson, M. (2020). googleLanguageR: Call Google's 'Natural Language' API, 'Cloud Translation' API, 'Cloud Speech' API and 'Cloud Text-to-Speech' API. R package version 0.3.0. https://CRAN.R-project.org/package=googleLanguageR.
Hai, R., Koutras, C., Quix, C., & Jarke, M. (2023). Data Lakes: A Survey of Functions and Systems. IEEE Transactions on Knowledge and Data Engineering.
Iansiti, M., & Lakhani, K. R. (2020). Competing in the age of AI: Strategy and leadership when algorithms and networks run the world. Harvard Business Press.
Lamarre, E., Smaje, K., & Zemmel, R. (2023). Rewired: the McKinsey guide to outcompeting in the age of digital and AI. John Wiley & Sons.
Nagle, T., Redman, T., & Sammon, D. (2020). Assessing data quality: A managerial call to action. Business Horizons, 63(3), 325–337.
Nargesian, F., Zhu, E., Miller, R. J., Pu, K. Q., & Arocena, P. C. (2019). Data lake management: challenges and opportunities. Proceedings of the VLDB Endowment, 12(12), 1986–1989.
Rigby, D., Elk, S., & Berez, S. (2020). Doing agile right: Transformation without chaos. Harvard Business Press.
Rogers, D. L. (2023). The Digital Transformation Roadmap: Rebuild Your Organization for Continuous Change. Columbia University Press.
Ross, J. W., Beath, C. M., & Mocker, M. (2019). Designed for digital: How to architect your business for sustained success. MIT Press.
Ruparelia, N. B. (2023). Cloud computing. MIT Press Essential Knowledge series. Revised and Updated Edition. MIT Press.
Schmarzo, B. (2016). Big Data MBA: Driving business strategies with data science. John Wiley & Sons.

Sutherland, J. (2014). Scrum: the art of doing twice the work in half the time. Crown Currency: New York, NY, USA.

Vaz, N. (2021). Digital business transformation: How established companies sustain competitive advantage from now to next. John Wiley & Sons.

Data Science in Organizations

4

This Chapter discusses data science in organizations. It starts by explaining the difference between data science and applied statistics, before focusing on the skills and workload of data scientists in industry. Data scientists engage in a broad range of activities: from forming hypotheses, choosing data, collecting data, all the way to communicating the results in non-technical terms to decision makers, and writing up evaluations of policies that were set up on the basis of their findings. The chapter then considers the leadership skills required of managers in digital organizations.

This chapter discusses an important premise for the book: managers need not necessarily be experts in the area of data science and AI, but they should have some awareness of data analysis. They need to be able to have meaningful discussions with their data scientists and quantitative analysts. Lacking a basic level of skills needed for such dialogues could be seen as a lack of commitment to the digital strategy, potentially making managers appear untrustworthy. Managers also need to have a strong strategic vision and leadership skills, and be willing to explore which technologies would fit well with their overall strategy. Abdication of leaderships skills to AI is a bad idea. In the digital age, leadership skills are more important than ever!

4.1 The Difference Between Data Science and Applied Statistics

In an academic research department, an applied statistician would be someone who focuses their research on a small range of techniques (regression models, time series models, etc.), and uses a small number of clean datasets. The interest is not so much in the data, or in the results, but in improvements in the theoretical properties of the techniques. Parts of the investigative process, such as problem analysis, measurement, and data cleaning, are given little attention (Wild and Pfannkuch 1999). Such an applied statistician has an interest in mathematical details, and may spend considerable energy on the formal mathematical proofs underlying statistical theorems. An applied statistician has a very different role to that of a corporate data scientist. While an applied statistician receives a clean dataset and applies statistical reasoning to it, data scientists work in the wild.

A data scientist working in a firm has a broader range of interests, and familiarity with a broader range of statistical techniques, because of the need to have "the right tool for the right job". Such a data scientist will also use a variety of software packages (R, Python, Matlab, etc.) depending on the task at hand. While academic researchers are often quite poor at coding and software development (Goble 2014), we would have higher expectations from a data scientist. Furthermore, data science requires a multidisciplinary approach, not only drawing on statistics, but also disciplines such as computer science, data mining, mathematics, and computer programming. In particular, Taddy (2019) considers that business data science draws most closely on the three disciplines of machine learning, economics, and statistics. Machine learning provides useful insights regarding how to automate and scale; Economics provides tools for causal inference and structural modelling; and statistics gives valuable insights regarding the uncertainty involved in the exploratory forward-looking predictive and prescriptive analyses of business data science.

Big data has enabled the rise of business data science. In the previous era of **Business Intelligence** (BI), firms focused on producing standardized backward-looking reports to describe performance outcomes in previous periods. These outputs were essentially aggregated tables of numbers, that were not affected by statistical uncertainty because the numbers came from highly structured data from previous periods (Schmarzo 2016). Data Science takes a different approach compared to BI. Data Science focuses on leveraging new sources of data (structured data as well as unstructured data such as text data) to obtain new variables and metrics, while using exploratory techniques, to search for new patterns that might yield useful forward-looking predictive and prescriptive analytics.

4.2 Skills of a Data Scientist

Data scientists use statistical methods to create knowledge from raw data. They use a large set of techniques, and a broad range of possible data sets, from traditional indicators (e.g. quarterly sales, broken down by region) to nonstandard data sources (e.g. text data from social media, data collected from Internet-of-Things (IoT) devices, user search data from Google Trends, information on competitor's discounted special offers, and information on the weather and local events). Data scientists also benefit from having a broad skill set, and a broad social network. In addition to knowledge of scientific methods, data scientists would benefit from a

4.2 Skills of a Data Scientist

broad array of soft skills, because of the need for tacit contextual knowledge and insider information when making sense of the signals emerging from various data sources.

Data scientists also need to have a deep knowledge of the business and nuanced insider knowledge of the many issues and tradeoffs that feed into decision-making. For example, it could be inappropriate if your data-driven recommendations emphasize a cost reduction plan, whereas the firm wants to focus instead on boosting sales. Data scientists should learn about the business, in terms of its strategic vision, its portfolio of potential new projects, its various long and short-term goals, its performance in terms of various indicators, and relevant information in sources such as annual reports. Data scientists can gain valuable contextual knowledge by "getting out there!" and touring the facilities, visiting customers, engaging in conversations at the water cooler, riding with the delivery drivers, and accompanying a service technician on one of their shifts (Kenett and Redman 2019).

Kenett and Redman (2019) present the life cycle view of data analytics in the following 8 steps, and emphasize that data scientists need to function at each of these 8 steps. Note the contrast with applied statisticians, whose scope might only cover the two central tasks of "Data Analysis" and "Formulation of Findings."

- PROBLEM ELICITATION: Observing and listening carefully, interacting with various stakeholders, asking probing questions, and trying to home in on the most important underlying issues.
- GOAL FORMULATION: Synthesizing various perspectives, seeing through the noise to clearly formulate the problem.
- DATA COLLECTION: Identify relevant data sources, while being aware of their context and limitations. Understanding the strengths and limitations in the data, dealing with data quality issues, and deciding when new data is needed.
- DATA ANALYSIS: Use a variety of methods to create insights from the data, using a broad range of techniques. Techniques include exploratory data visualization, descriptive statistics, causal modelling, predictive models, and data-based prescriptive recommendations for action.
- FORMULATION OF FINDINGS: Converting the results from the data analysis to actionable recommendations communicated clearly to decision-makers. This may require different reports tailored for different audiences (senior managers, specialist departments, etc.).
- OPERATIONALIZATION OF FINDINGS: Setting out a roadmap for implementation of the data-based recommendations, with suggestions for who does what, when, and how. This involves answering questions from stakeholders, and updating the plan as new data appears. This may also involve cautiously advising on situations that extend beyond the scope of the original data analysis.
- COMMUNICATION OF FINDINGS: Communicating the findings and their implications to all stakeholders who may be impacted, going beyond the group of individuals who are actually involved in taking action to implement the recommendations. Making clear where the data ends and intuition must take over.

- IMPACT ASSESSMENT: Setting up and executing a strategy for evaluating the impact of the initiatives and actions taken. Ideally, such an impact assessment will involve objectively-measured quantitative data and will aim for estimations of causal effects (e.g. A/B testing) rather than statistical associations (see Chap. 12).

While data scientists play a crucial role in digital transformation, effective leadership is also essential for success, as leaders need to provide a clear vision, earn trust, and motivate employees to embrace new technologies and ways of working.

Data scientists can be said to possess 5 traits (Davenport 2014, p. 88): they are Hackers, Scientists, Trusted Advisers, Quantitative Analysts, and Business Experts. Going more in detail:

- **Hacker**: Proficient in coding and understanding the architectures of big data technology.
- **Scientist**: Curious and interested in exploring new hypotheses, oriented towards taking action, and engaging in evidence-based decision making. A relentless focus on discovering the real problems, and learning about the world.
- **Trusted Adviser**: Excellent inter-personal communication skills, and the ability to understand decision processes and frame decisions. Data scientists know that they will not receive opportunities to contribute to important decisions unless they are trusted.
- **Quantitative Analyst**: Expertise in data visualization, statistical analysis, and machine learning algorithms, including skills for the analysis of unstructured data (such as text, images, and video).
- **Business Expert**: Deep understanding of how the business makes money, and a sense of which tasks could benefit from the application of analytics and data science. Data scientists discover the real problems that may not be well-articulated by managers (Kenett and Redman 2019). When managers admit that something "just doesn't feel right", data scientists know that they are on to something, and can sniff out the underlying issues. Data scientists are also wary of decision-makers who attempt to use data science to justify decisions they've already made.

4.3 Skills of Leaders for Digital Transformation

The explosion of data is a daunting challenge for decision-making. The ultimate goal is to collect and synthesize information in order to put in place an evidence-based decision process powered by analytics. Data science and AI can help avoid bad decisions that come from "gut" decision-making, decision fatigue, and decision biases such as confirmation bias (seeking evidence that confirms your initial position, rather than being open to the possibility of disconfirming evidence). However, the sheer amount of data available can lead to distractions. Unless the use of

4.3 Skills of Leaders for Digital Transformation

data is clearly linked to relevant hypotheses and objectives, the data may end up guiding the analysis, rather than the business goals guiding the data and analytics.

There is a tendency for some leaders to abdicate their leadership to data scientists and technicians (De Cremer 2024). Leaders may say things such as: "Our new strategy is to invest aggressively in big data and AI, and therefore we will wait for the tech guys to tell us what to do next." This reasoning is flawed. Digital transformation is fundamentally not about technology, but about strategy and leadership (Rogers 2016, p. 239). Technology can become a distraction from strategy, and technology should only be used when it is consistent with attaining the firm's strategic objectives. The root causes of business failures are typically to be found in the domain of management, rather than being due to technology. Digitalized firms may excel at turning data into information and better decisions, but these decisions need to be linked to achieving the overall business goals. The problem is not that leaders need to step back and follow the technicians, but that leaders need to start leading. In the digital age, leadership matters more than ever. Furthermore, many of the required leadership skills are traditional, classic leadership skills (such as flexibility, earning trust, and motivating others around a common vision) that make sense even when the word "digital" is not used (Kane et al. 2019).

Leaders in the digital age need to have a strong vision, and sense of purpose, and they draw new technologies into their strategic vision. Leaders are not swayed by the latest fads and trends, but are only interested in new technologies so far as they align with their vision. As such, leaders decide which questions are strategically important, they do not delegate such tasks to the technology experts.

Successful leaders in the digital age will have a strong sense of the purpose of the organization, that is clearly demonstrated in their actions and decisions. This sense of corporate purpose will enable clear communications during meetings with data scientists, helping to inform data scientists what the priorities are.

Effective leaders will need to be perceived as honest and trustworthy on the part of employees, to avoid the situation where employees lack motivation, or even attempt to sabotage, initiatives for digital transformation and the application of AI. Leaders will need to generate enthusiasm among employees for how new technologies can help to achieve the strategic business purpose. Leaders will also need to appear to be competent and genuinely interested when discussing technical subjects with data scientists and AI technicians. It can be very demotivating for a technical team if a top manager says that they don't understand what they are talking about, and have to summarize their project in 3 short bulletpoints (De Cremer 2024). Top managers are expected to invest serious efforts in being prepared for digital transformation plans: "You should plan on having each top executive invest a minimum of 20 h of learning before they can be ready to productively engage in defining a digital roadmap with their colleagues." (Lamarre et al. 2023, p. 22).

So, what kind of skills are needed by managers? Not everyone needs to be a data scientist, of course. Managers don't necessarily need to learn how to code. However, managers need to think like data scientists, while having a firm grasp of the company's overall strategic mission.

A crucial skill is being able to engage in conversations with technical staff. On the one hand, managers need to have a sufficient level of understanding of data science and AI topics that they can understand the topic well enough to allow for meaningful exchanges, explain how digital initiatives relate to the core business, and also learn about the constraints and issues affecting digital activities. The popular press often refers to the shortage of Data Scientists, but there is also a shortage of digitally-literate managers who possess the valuable "**fusion skills**" that allow them to recognize valuable opportunities at the intersection of digital technologies and business priorities. Managers should stay curious and seek to continually update their skills in this fast-paced area. On the other hand, managers will certainly find themselves in situations where they do not understand what their technicians are talking about. Technicians may even deliberately use overcomplicated language from time to time, or pitch ideas in ways such as "I don't want to explain how it works, but it's AI" (HBR 2023). In these cases, it is important to persevere and keep probing the other party with tough questions to get a better grasp of the main issues (Davenport 2013). Blindly trusting in the perceived genius of quantitative analysts is not a good leadership strategy for successful digital transformation. Examples of such questions that managers could pose to their "quants" include:

- What data is going into the analysis, and how representative is this data?
- Might the results be sensitive to the modelling assumptions?
- What results are found using alternative techniques, including more transparent techniques?
- Are there any risks (such as ethical risks) that could potentially arise from this model?

All the while, it is worth keeping in mind that there is no "magic", and that even complicated concepts can eventually be explained in clear language. Managers can also ask their analysts thought-provoking questions relating to how the technical analysis relates to the fundamental strategic priorities of the business. Data-analytic thinking also enables managers to better assess the value of ideas proposed by employees, consultants, or potential partners.

Soft skills, such as listening and empathy, being authentic, and using a rhetoric of collaboration, are also important leadership skills for managers. Far too often, top managers talk about their lofty plans to become digital companies (regardless of which sector they come from), to automate as much as possible, and to replace human workers with robots and AI, and all of this has the unhealthy effect of making workers worry for their jobs. Workers may be unenthusiastic, potentially sabotaging new digital initiatives through passive resistance, which has led many digital initiatives to have low take-up and unimpressive performance, and ultimately end up as expensive mistakes. As such, it is important to take a human-centered approach to AI adoption, to be aware of the motivations and fears of the various stakeholders, and to practice emotional intelligence (De Cremer 2024). Managers should invest in earning the trust of employees, and managers should also reassure employees with statements that data science and AI are tools to augment the capabilities of workers, rather than to automate (and hence replace) worker tasks. Indeed, the emphasis should be on using digital technologies to augment your workers' capabilities, and to unlock more innovation and creativity in workers, rather than to automate away their jobs. This will require effective communication skills, empathy, and a spirit of collaboration, that go beyond technical details in order to connect with the core values of the organization.

Leaders will need to convince their employees to buy in to digital initiatives, and to become convinced that making efforts to advance the digital transformation will be worth their while. Leaders will need to listen to employees' questions, understand employees' perspectives, and explain the risk and benefits that digital transformation and AI adoption will bring for them. Leaders should consult with a wide variety of employees, from different departments, in part because a failure to consult with certain groups could leave them to lack involvement and commitment to the initiative.

In addition to leadership skills, however, organizational culture is also crucial for digital transformation, as it shapes the behaviors, attitudes, and values of employees and influences their willingness to embrace change and innovation. We now turn to organizational culture.

4.4 Organizational Culture

The pursuit of digital transformation has implications for the appropriate culture. Digital firms seek the productivity gains that come from superior decision-making that is based on insights from data. As such, decisions are less influenced by the authority figures in traditional hierarchies, such as the "**HIPPO**" (i.e. the Highest Paid Person's Opinion; Schmarzo 2016). Instead, decisions are made based on high-quality data, wherever it may come from. Jeff Bezos famously summarized this idea:

> If this is a decision based on opinions, then my opinion wins. However, data beats opinion. So bring data[1]

The emphasis on data as the foundation for decision-making means that the traditional top-down hierarchy is upended. The primacy of data means that, in a digital world, good ideas can come from anywhere, and as a result, new communication channels need to be set up, where ideas can flow from low-level employees up to the top, and from one functional division to another.

An inspiring story regarding an organizational culture that encourages innovation comes from the case of Greg Linden, a former developer at Amazon.[2] Greg

---

[1] Jeff Bezos, quoted in Hoffmann and Yeh (2018, p. 165).
[2] See https://glinden.blogspot.com/2006/04/early-amazon-shopping-cart.html [last accessed 16 Nov 2024], and also Rogers (2023).

was working in the area of the customer checkout process, when he had the idea to offer shoppers a final set of recommendations at the time of check out, based on the items that were in their cart. The senior managers hated the idea, which seemed to break a golden rule of e-commerce: do not distract the shopper once they have started the checkout process. Greg, however, started from the intuition that check-out shelves in real-world supermarkets often present last-minute purchase options (such as candy and impulse buys) to customers approaching the checkout. Greg was explicitly forbidden to work on the project… but he hacked up a quick prototype version on a test site anyway. This prototype received opposition from a marketing senior vice-president who rallied others to the cause of killing the initiative. Greg was told that he was forbidden to work on this any further. Undeterred, however, he prepared the feature for an online test, to gather data to measure the sales impact. Indeed, the culture at Amazon at the time was such that it was hard for even a top executive to block a test experiment. The data came back positive, and resources were immediately applied to developing and launching a full version of the shopping cart recommendations feature.

This case can be considered as an example of best-practice. Too few organizations have such a culture that puts data and evidence over the opinions and hunches of top managers.

Digital organizations therefore need to do more than just append some data analysis to existing activities. They need to intentionally rethink their hierarchies, and encourage more collaborative interactions. Everyone in the organization can potentially be the source of an important new idea. Digital transformation entails a broad cultural transformation away from a traditional hierarchy, to strive for a more collaborative, risk embracing, and agile culture, in order to get the most out of talented employees (Kane et al. 2019).

Digital firms should strive to ensure that the workplace is a safe space for employees to share ideas. Group brainstorming meetings can give managers useful insights into how new technologies are being used, and concerns about new digital initiatives. In such meetings, managers should not speak first, but should try to learn. If the workplace is perceived as a supportive environment, then employees will be more enthusiastic about digital initiatives, maybe even suggesting their own initiatives for adopting new technologies for specific tasks to boost productivity and lower the error rates. Instead, if the workplace operates according to a top-down imposition of new technologies, employees might be unenthusiastic, and harbour resentment towards these initiatives, preferring not to take up the new digital tools, and in some cases sabotage these initiatives in secret. AI processes can easily be sabotaged by micromanaging and intervening in the AI process rather than taking AI insights as an imperfect but potentially valuable complement to decision-making (De Cremer 2024). Therefore, it is important to inspire employees to genuinely collaborate, rather than pressuring them to agree to new initiatives. Instead of forcing change onto employees, it is better to focus on creating conditions where transformation is ripe to occur (Kane et al. 2019). After all, culture can be described as what employees do when no-one is looking (Rogers 2023).

The role of culture again highlights the importance of leadership. Leaders should have a clear vision that aligns with the corporate purpose, and leaders should clearly communicate this vision to employees and stakeholders, to convince them of the advantages, to learn about their concerns, and to seek strategic alignment towards the new goal. New initiatives will not succeed if they are forced, but they should be implemented based on trust and respect.

Consider the following two perspectives. First, a manager instructs employees that they have to start using a new labour-saving digital tool, and when asked how their jobs will be affected, replies that this decision will be made by the AI department and the digital transformation committee. Second, a manager introduces a new labour-saving digital tool, after having discussed it with employees during a long process that led to some unexpected improvements in the tool, thanks to feedbacks from the employees. When asked by employees how their jobs will be affected, the manager replies that this tool is designed to augment humans rather than automate humans away, and in any case could not ever completely replace humans because it can only replace a narrow range of tasks. The manager refers to the corporate mission to explain that everyone's job tasks will change in the future, but that through collaboration, flexibility, and agility, there will always be a place in the firm for those who want to keep contributing to the corporate mission. The second of these two perspectives will no doubt be more successful in their digital transformation efforts.

Further Reading

Kenett and Redman (2019) discuss the role of Data Scientists, and how they differ from Applied Statisticians. Rogers (2023, especially Chap. 7) discusses organizational culture in digitalized firms. De Cremer (2024) focuses on the leadership skills that managers need in the age of AI.

References

Davenport, T. (2014). Big data at work: dispelling the myths, uncovering the opportunities. Harvard Business Review Press: Cambridge, MA.
Davenport, T. H. (2013). Keep up with your quants. Harvard Business Review, 91(7/8), 120–123.
De Cremer, D. (2024). The AI-savvy leader: Nine ways to take back control and make AI work. Harvard Business Press: Cambridge, MA.
Goble, C. (2014). Better software, better research. IEEE Internet Computing, 18(5), 4–8.
HBR. (2023). HBR Guide to AI Basics for Managers. Harvard Business Review Press. Massachusetts: USA.
Hoffman, R., & Yeh, C. (2018). Blitzscaling: The lightning-fast path to building massively valuable companies. Currency Books, New York, USA.
Kane, G. C., Phillips, N., Copulsky, J. R., & Andrus, G. R. (2019). The Technology Fallacy: How people are the real thing to digital transformation. MIT Press.
Kenett, R. S., & Redman, T. C. (2019). The Real Work of Data Science: Turning data into information, better decisions, and stronger organizations. John Wiley & Sons.

Lamarre, E., Smaje, K., & Zemmel, R. (2023). Rewired: the McKinsey guide to outcompeting in the age of digital and AI. John Wiley & Sons.
Rogers, D. L. (2016). The digital transformation playbook: Rethink your business for the digital age. Columbia University Press.
Rogers, D. L. (2023). The Digital Transformation Roadmap: Rebuild Your Organization for Continuous Change. Columbia University Press.
Schmarzo, B. (2016). Big Data MBA: Driving business strategies with data science. John Wiley & Sons.
Taddy, M. (2019). Business data science: Combining machine learning and economics to optimize, automate, and accelerate business decisions. McGraw Hill Professional.

Statistical Associations Using Regression

5.1 Univariate Distributions

Analysis of distributions, one variable at a time, can be a useful exercise. Figure 5.1 (left) shows the distribution of log sales, using a **histogram** (with blue rectangular bars, to show the frequency in each size class) and also a **kernel density** plot in pink. A histogram can be considered to be a rectangular approximation to a distribution, and the shape of a histogram changes depending on the number of bins, and the class sizes. A kernel density is a smoothed histogram, and (according to a rough intuition, Wickham et al. 2023) it is similar to what we would get if we could lay a piece of cooked spaghetti on top of a histogram. A kernel density comes with a smoothing parameter, that can be modified to vary the shape from extremely smooth to extremely jagged (picking up all the bumps in the data). The kernel densities in Fig. 5.1 (left and right) use the default settings.

The distribution in Fig. 5.1 (left) is **skewed** to the right: it looks as if we grabbed the corner of the data and pulled it out to the right. Skewed to the left would be the mirror-image: as if the distribution was stretched out to the left. With a right-skewed distribution, most of the data points are at the lower end of the range of the x-axis, but there are a small number of extreme values at the upper end. In our case, most firms have a relatively low value of net sales, but a small number of firms have extremely large values. Indeed, the firm size distribution is typically skewed, with some large firms being thousands of times larger than other firms. In fact, Fig. 5.1 (left) does not show the full range of the sales variable, because if we showed the full range, then most of the action would be squashed down even further at the left end of the figure, and it would be hard to see what is happening.

Supplementary Information The online version contains supplementary material available at https://doi.org/10.1007/978-981-95-2433-4_5.

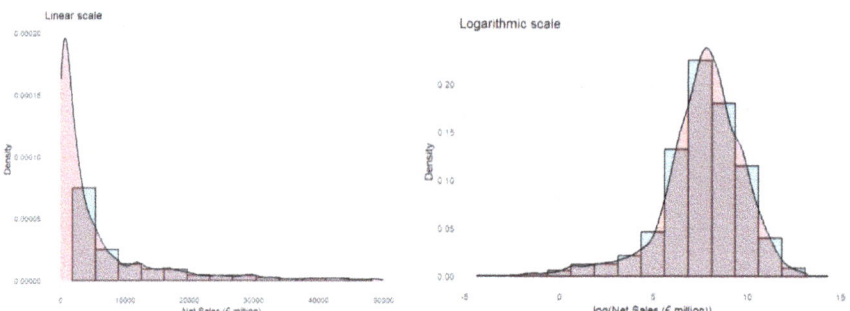

Fig. 5.1 Univariate analysis: histogram and kernel density plot. (*Source* Author's elaboration. See R code)

Skewed data can be hard to plot on graphs, and can cause problems for some quantitative techniques (such as OLS regression) which work better with normally distributed variables.

Skewed data can lead to confusion in cases where the skew is not fully appreciated. Consider a sample of 10 firms, 9 of which are aged 5 years, and one is aged 100 years. This is not an unrealistic scenario for firm ages. The average age would be 14.5, which does not correspond well to any of the firms. Instead, with skewed distributions, the median could be a more informative indicator of central tendency (here, the median age would be 5 years).

Figure 5.1 (right) therefore transforms the data by taking the logarithm of sales. Taking logarithmic transformations is not an "artificial" or "deceptive" way of handling data, instead it is a very common procedure in data science, that you should not hesitate to apply.

5.2 Correlations

Correlation refers to the degree to which two variables are related. Correlations are quantified in terms of correlation coefficients, which can vary from −1 (perfect negative relationship) to zero (no relationship) to +1 (perfect positive relationship). Figure 5.2 presents some examples of clouds of datapoints, alongside the corresponding correlation coefficient. The top row of Fig. 5.2 shows that noisy relationships between variables have lower correlation coefficients. The middle row of Fig. 5.2 shows the correlation coefficient is perfectly positive (or perfectly negative) if all the datapoints are positioned on the same line of positive slope (or negative slope, respectively). Interestingly, the central case in the middle row corresponds to a correlation of zero, even though the datapoints are on the same straight line, because the value of y is the same, no matter the value of x. The third row of Fig. 5.2 illustrates a weakness of the correlation coefficient: it can only detect linear relationships between variables, and cannot detect nonlinear patterns.

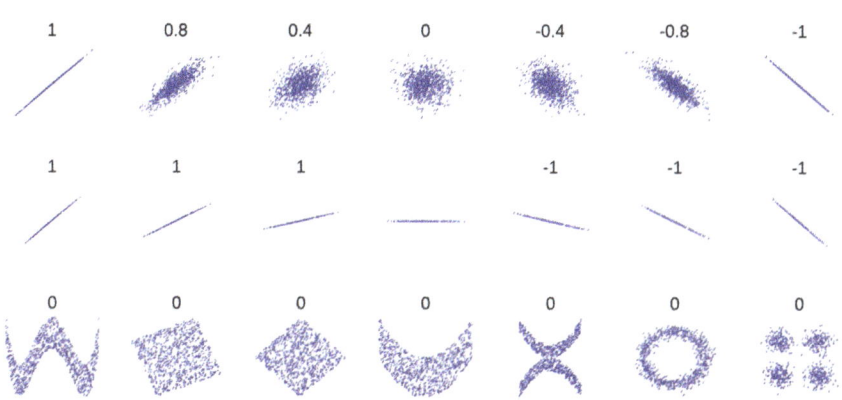

Fig. 5.2 Bivariate clouds of datapoints and the corresponding correlation coefficients. (*Source* DenisBoigelot, original uploader was Imagecreator, CC0, via Wikimedia Commons. https://commons.wikimedia.org/wiki/File:Correlation_examples2.svg)

For the sake of completeness, we can show the equation for the correlation coefficient between X and Y, which is $\rho_{X,Y}$:

$$\rho_{X,Y} = \frac{E[(X - \mu_X)(Y - \mu_Y)]}{\sigma_X \sigma_Y}$$

where μ is the mean, σ is the standard deviation, and E[] is the expectation. Looking at this equation, we can see that there is a positive correlation when above-average values of x occur alongside above-average values of y. Contrariwise, there is a negative correlation when above-average values of x occur alongside below-average values of y.

Scatterplots (and **scatterplot matrices**) are useful tools to inspect how two variables are correlated. Scatterplots are widely used in data exploration because one can instantly see how variables are related, if the relationship is linear or nonlinear (e.g. U-shaped), and whether the datapoints are evenly distributed across the range, or bunched up in a corner. For an example of a scatterplot matrix on R's inbuilt "iris" dataset, type "`pairs(iris)`" into R, and marvel how such a rich output comes from typing just 11 characters.

While correlations provide a useful way to understand the relationship between two variables, OLS linear regression offers a more powerful and flexible tool for modeling and predicting the relationship between a dependent variable and one or more independent variables.

5.3 OLS Regression

Ordinary Least Squares (OLS) linear regression is the benchmark regression model, and the starting point for a large group of other regression models. OLS regression seeks to fit a straight line through a cloud of datapoints, in order to summarize the main trends and patterns in the data.

The OLS regression equation can be written as follows:

$$y_i = \alpha + \beta x_i + \varepsilon_i$$

Our **dependent variable** (i.e. the outcome we want to explain) is y, and we also have one or more **explanatory variables**. OLS regression is more useful in cases where we have many explanatory variables, but in this simple example (in Fig. 5.3) we only have one explanatory variable, x. α is the intercept term, which is the expected value of y when x equals zero. β is the regression coefficient, which informs us of the relationship between x and y. If we increase x by one unit, then y increases by β units. Finally, ε_i is the residual error term for observation i, which is that part of the outcome in y that is not explained well by the variation in x. OLS tries to find the best-fitting line, such that the error terms ε_i are as small as possible. The OLS best-fit line always runs through the point $(\bar{x}, \bar{y})$, which is the average of both variables (Imai and Williams 2022).

Figure 5.3 gives a simple introduction. The raw datapoints are represented by the black circles. The OLS regression line is the orange line. The OLS regression line corresponds to the OLS predictions for the datapoints, hence the orange triangles are the predicted values that correspond to the actual values of the datapoints (represented by the black circles). The distance between the true values of the data points, and the predicted values of the datapoints (orange triangles) is represented by the vertical blue dashed lines, which indicate the error of the OLS model. The OLS line of best fit is the line that minimizes the lengths of the blue lines (to be precise, OLS minimizes the sum of the lengths of the blue lines squared).

Figure 5.4 shows the OLS regression output in R. There is a lot of information that we don't need right now. Of primary interest are the estimated **coefficients**. The **intercept** term is 0.8104. This means that when x = 0, the regression model predicts that y = 0.8104. The coefficient on x is 1.4198, which suggests that the relationship between x and y is positive, and that each increase in x by one unit is associated with an increase in y by 1.4198 units. The coefficient is statistically significant, and we can see this by looking at the **t statistic** (here called "t value") which is relatively large at 4.020 (larger than the usual reference point of about 2). Is the coefficient estimate of 1.4198 truly different from zero, or could it be just due to chance? The statistical significance of the coefficient estimate is evaluated with respect to the baseline reference case (or null hypothesis) that it is equal to precisely zero.

The null hypothesis is that the regression coefficient is equal to zero. The **p-value** assumes that the null hypothesis is true, and is a measure of the strength of the evidence against the null hypothesis. The p-value (here called "Pr(>|t|)",

5.3 OLS Regression

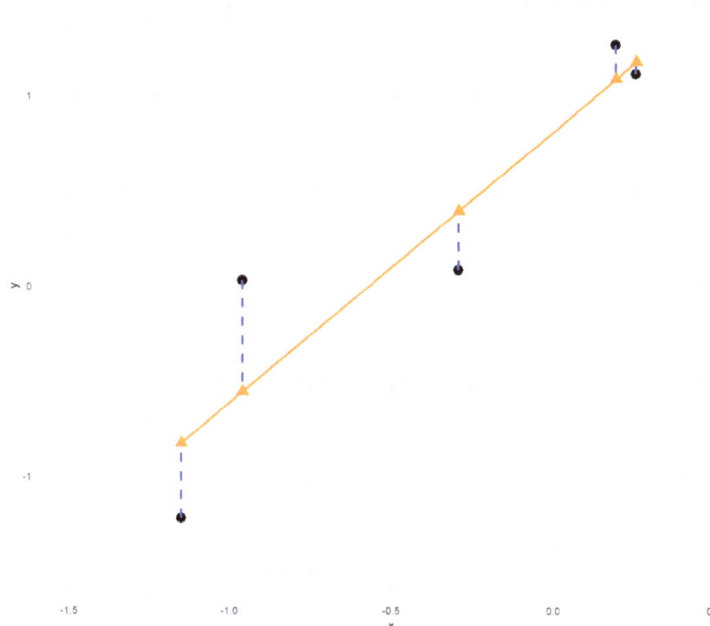

Fig. 5.3 OLS regression. The graph shows 5 raw datapoints (black circles), OLS line of best fit (orange line), OLS predicted values (orange triangles), and the error terms (blue dashed lines). (*Source* Author's elaboration, see R code for details)

```
> summary(OLSmodel)

Call:
lm(formula = y ~ x, data = data1)

Residuals:
       1        2        3        4        5
 0.58544 -0.30968 -0.06123 -0.39350  0.17898

Coefficients:
            Estimate Std. Error t value Pr(>|t|)
(Intercept)   0.8104     0.2469   3.282   0.0463 *
x             1.4198     0.3532   4.020   0.0276 *
---
Signif. codes:  0 '***' 0.001 '**' 0.01 '*' 0.05 '.' 0.1 ' ' 1

Residual standard error: 0.458 on 3 degrees of freedom
Multiple R-squared:  0.8434,    Adjusted R-squared:  0.7912
F-statistic: 16.16 on 1 and 3 DF,  p-value: 0.02765
```

Fig. 5.4 OLS regression output in R. (*Notes* Author's elaboration, see R code file for details)

and taking the value 0.0276) is the probability that the t-statistic would be at least as large in magnitude as its observed value, if the true value of the regression coefficient were zero (Greenland et al. 2016). The p-value here is 2.76%, which is considered to be just too unlikely (if we take the usual 5% significance level), and so we would prefer not to accept the null hypothesis, and to tentatively conclude instead that the coefficient is significantly different from zero.

Another way of checking the statistical significance of the regression coefficients is to look for significance stars. In the case of the intercept term, and also for x, we see a single significance star "*" at the end of the row, which (according to the legend in the results output) suggests that these coefficients are statistically significant at the 5% level.

We might also be interested in the R^2 **statistic**, which we would expect to be in the range from 0 to 1. Values of the R^2 statistic close to zero suggest that the OLS regression model has a poor explanatory power, whereas values of the R^2 statistic close to one indicate a near-perfect fit. In our case, the R^2 statistic is 0.8434, which indicates a good fit.

The R^2 statistic is calculated in terms of the sum of squared residuals (SS_R) and the total sum of squares (SS_T):

$$R^2 = 1 - \frac{SS_R}{SS_T}$$

The R^2 statistic is close to one in cases where the sum of squared residuals is relatively small (i.e. adding up the squared values of the vertical dashed blue lines in Fig. 5.3) compared to the total sum of squares (what the squared residuals would be if we left aside our OLS model and simply took the average value of y as our best prediction of y).[1]

5.4 Logistic Regression

OLS regression is a linear regression technique, that basically tries to fit a straight line through the cloud of points. OLS works well in the case of continuous variables, but performs less well in the case of binary variables. This can be a problem, because in the real world many outcomes are binary, such as:

- Will this person pay their bills, or default?
- Is this a good or a bad product review?
- Will the Tokyo Yomiuri Giants win or lose this game?

---

[1] A nice intuition can be found at https://seeing-theory.brown.edu/regression-analysis/index.html [last accessed 15th July 2025].

5.4 Logistic Regression

With a binary outcome variable, the conditional mean of the outcome y (conditional on the explanatory variables x) becomes:

$$E[y|x] = p(y=1|x) \times 1 + p(y=0|x) \times 0 = p(y=1|x)$$

As such, we are modelling a probability that $y = 1$. We therefore need a "link function" that yields values for the probability that is between 0 and 1.

$$p(y=1|x) = f(\beta_0 + x_1\beta_1 + \cdots + x_p\beta_p)$$

The function f() takes information from the explanatory variables x, and gives a predicted probability for the outcome y. No matter whether the values for x are extremely large or extremely small, the predicted probability for y is in the range [0, 1]. Specifically, the function f() used by **logistic regression** is the logit link function:

$$p(y=1|x) = \frac{e^{x'\beta}}{1+e^{x'\beta}} = \frac{\exp[\beta_0 + x_1\beta_1 + \cdots + x_p\beta_p]}{1+\exp[\beta_0 + x_1\beta_1 + \cdots + x_p\beta_p]}$$

where e is Euler's number (i.e. the base of the natural logarithm; 2.71828).

This can be rearranged to help interpret the β coefficients:

$$\log\left[\frac{p}{1-p}\right] = \beta_0 + x_1\beta_1 + \cdots + x_p\beta_p$$

where p denotes $p(y=1|x)$.

Hence, we can consider that logistic regression is a linear model for the log odds, $\log[p/1-p]$.

What does "**log odds**" mean? Let's start by considering the statistical notion of odds: the probability that an event happens, divided by the probability does it does not happen. If an event has a probability of $1/4$, then the odds would be $\frac{1/4}{3/4} = 1/3$. Table 5.1 shows some examples of the probability p, the corresponding odds, and the log odds. If the probability lies between 0 and 1, the odds is always positive, and the log(odds) ranges from $-\infty$ to $+\infty$. If we are comfortable in working with the log odds, then we can consider that logistic regression is a linear model for predicting the log odds.

Logistic regression is similar to OLS regression in a number of ways, but there are also many differences. Logistic regression does not give an R^2 statistic in the same way as what we get in the OLS regression output, although there are some alternative indicators of model fit for logistic regression, such as the **deviance**, the **Akaike Information Criterion (AIC)**, and the **Nagelkerke R^2 statistic**.

Logistic regression is therefore more appropriate for the case of a binary dependent variable. Logistic regression is widely used and is sometimes called a machine learning technique. While OLS tries to fit a straight line through the datapoints, logistic regression fits a logistic S-curve through the datapoints, as will be illustrated in the example in the next section.

Table 5.1 Examples of values for the probability, odds, and log odds

Probability	Corresponding odds	Logarithm of the odds
0.01	$1/99 = 0.0101$	-4.595
0.1	$10/90 = 0.111$	-2.197
0.5	$50/50 = 1$	0
0.9	$90/10 = 9$	2.197
0.999	$999/1 = 999$	6.907

Source Author's elaboration, in the style of Provost and Fawcett (2013, p. 97)

5.5 R Example: OLS and Logistic Regression

This example focuses on simulated data regarding the relationship between a quantitative indicator of effort, and the test score obtained, for a group of 12 students. To begin with, we generate our simulated data, keeping the dataset small (with 12 observations) to make it easier to see what is happening.

Looking at the datapoints, there seems to be a positive relationship: higher effort is associated with higher scores. OLS regression finds the best-fit straight line that summarizes the relationship. We can say that all the predicted values of the datapoints lie on the line of best fit, and they are represented in Fig. 5.5 by the red triangles.

Looking at the OLS regression results table in R (not shown here), we see that the coefficient on effort is positive, at 3.483. This coefficient is statistically significant: the t-statistic is 2.758, and the p-value is 0.0202, far lower than the threshold of 0.05. We might also be interested in the indicators of model fit: the R^2 statistic is 0.4321, and the adjusted R^2 statistic is 0.3753. The way to interpret this regression coefficient of 3.483 is that, each time that effort increases by 1 unit, then the test score is expected to increase by 3.483. No matter whether you are at the bottom of the range of effort, or the top, there is a constant relationship between effort and test score. The straight line means that, wherever you are along the range, if effort increases by 1 unit, then we would expect an increase in test score by the same amount, which is 3.483 units. This property of a constant relationship between effort and test score is a useful property of OLS linear regression, because it is easy to interpret the effect sizes. For logistic regression, the interpretation is more complicated, as we will see.

When effort equals zero, test score is predicted to be equal to the intercept term, which is 59.448. When effort equals 1, the test score increases by 3.483 units, hence the test score is predicted to be about 62.9.

Now we move to model 2, where we have a binary variable (pass or fail). A test score of 60 or more is a pass, and a test score below 60 is a fail. OLS regression works less well with dependent variables that are binary. Logistic regression, however, is designed for cases where the dependent variable is binary. Figure 5.6 (left) shows the case of model 2, where we fit an OLS regression to the case of the

5.5 R Example: OLS and Logistic Regression

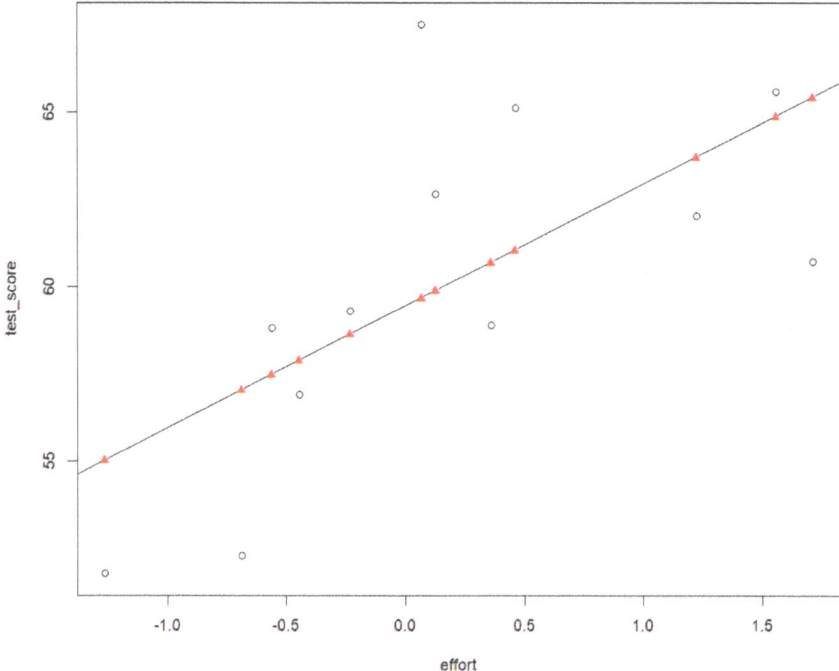

Fig. 5.5 OLS line of best fit, for the relation between effort and test score. (*Notes* Author's elaboration, see R code file for details)

binary dependent variable. The datapoints do not seem to line up well on OLS's straight line. The red triangles correspond to the theoretical predictions from the OLS model. At the lower and upper extremes of effort, we can even see some nonsense predictions. The predicted probability of a pass is negative at the lower end of effort, which technically makes no sense, because a probability should not be negative. At the upper end of the range of effort, some predicted probabilities are greater than 1, which is also a problem. Ideally, the predicted probabilities of a pass should all lie between zero and 1.

Now we move to model 3 in the R code, which is estimated using logistic regression. Logistic regression is similar to OLS regression, but it takes a different approach because it does not try to fit a straight line, but a logistic curve (shown on Fig. 5.6, right). As such, logistic regression avoids the problem of making predictions that are below 0% and greater than 100%. However, this comes at the cost of having a best-fit line that is not a straight line but a curve. The red line corresponds to logistic regression's predicted probabilities of a pass. With OLS regression and its line of best fit which is straight, every time the explanatory variable (effort) increases by 1 unit, there is the same expected change in the outcome. With logistic regression, however, the relationship between effort and

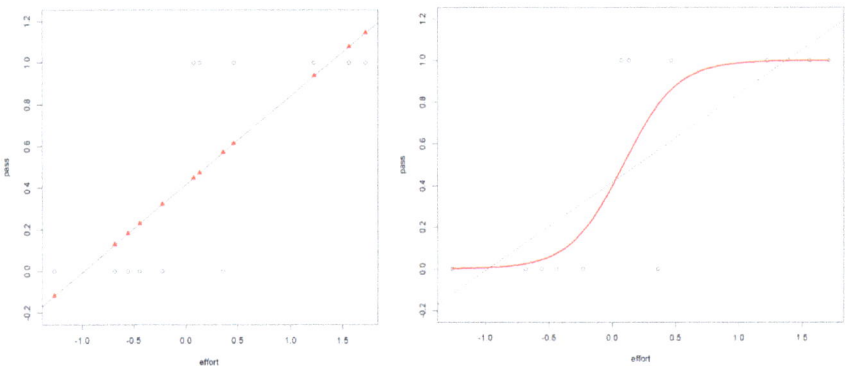

Fig. 5.6 OLS (left) and logistic regression (right) models for the case of the binary outcome variable. (*Notes* Author's elaboration, see R code file for details)

pass is nonlinear. For a one unit change in effort, the corresponding change in the probability of having a pass is not constant, but depends where you are on the range of the variable "effort". This makes logistic regression more difficult to interpret.

Model 3, estimated using logistic regression, shows a regression coefficient of 4.7923. This coefficient is not actually statistically significant at conventional levels, because the p-value is 0.119, and is above the usual threshold of 0.05. This is no doubt because our sample size is quite small. Instead, we could consider that this coefficient is statistically significant at the 15% level. Hence, how should we interpret the coefficient? We could plug our estimates of alpha (-0.4072) and beta (4.7923) into the estimated logistic regression equation:

$$p(y = 1) = \frac{exp[\alpha + \beta x]}{1 + exp[\alpha + \beta x]}$$

When $x = 0$, then we get a predicted probability of 0.40 for y.

When $x = 0.5$, we get a predicted probability of 0.88 for y.

And when $x = 1$, we get a predicted probability of 0.99 for y.

Note that we make the same change to x, adding 0.5 each time, and get very different outcomes for y. We refer to this as non-constant marginal effects of x on y. Logistic regression does not have a straightforward linear relationship between inputs and outputs, like for OLS linear regression.

Logistic regression also differs from OLS in terms of the model goodness of fit. OLS gives us the well-known R^2 statistic, but logistic regression does not do this. One idea could be to look at the AIC (Akaike Information Criterion) that comes as part of the logistic regression output. Another idea would be to use an alternative R^2 statistic, such as the Nagelkerke R^2 statistic. This is not part of the standard output for logistic regression, but it can be calculated quite easily (the R code shows it to be 0.748).

There are different opinions among econometricians regarding OLS and logistic regression. OLS can be applied to cases of a binary dependent variable, and this is referred to as **OLS-LPM: Linear Probability Model**. OLS has the advantage that the results are easier to interpret, in terms of the marginal effects: how a change in x is associated with a change in y. If x increases by one unit, how is y expected to change. A drawback of OLS is that it can potentially generate nonsense predictions, such as negative probabilities, or probabilities above 100%. But if we are not too concerned about these cases where OLS can sometimes generate nonsense predictions, then we might prefer to keep using OLS (Angrist and Pischke, 2008).

5.6 Lasso Regression

Lasso regression is a powerful tool for applying regressions to big data contexts. Lasso stands for Least Absolute Shrinkage and Selection Operator (Tibshirani, 1996). This section on Lasso regression starts by thinking about signal and noise. We then look at the theory of Lasso regression, before estimating a Lasso regression in R.

5.6.1 Overfitting

Statistical data are measurements that are hopefully accurate, but they may contain **noise**. High quality data has a high ratio of signal to noise. **Signal** corresponds to the valuable part of statistical information. Signal is the persistent pattern that is stable across different samples of the same process. Noise, in a regression context, corresponds to the unexplained residuals, or error terms, in regressions such as OLS linear regression. Noise can also refer to spurious correlations across variables, which are non-zero correlations that have no clear interpretation. Noise can correspond to the spurious variables in a regression model that might be statistically significant in one particular data sample, but are not likely to be statistically significant in other data samples. Noise leads to situations of false positives and false negatives. A **false positive** would be a variable that looks statistically significant in one regression model, but it should not really be considered to be a useful predictor. A **false negative** would be a variable that is not statistically significant in some regression output, but in fact it should be included in the set of important predictor variables.

The graph on the left of Fig. 5.7 shows the case of applying a linear regression to data that is clearly non-linear. The raw data seems to display a U-shape. The graph on the left is the case of underfit. Forcing the regression line to be a straight line means that it does not detect the nonlinear relationship. A more complex model (i.e. a nonlinear model) would do a much better job of fitting the data. In

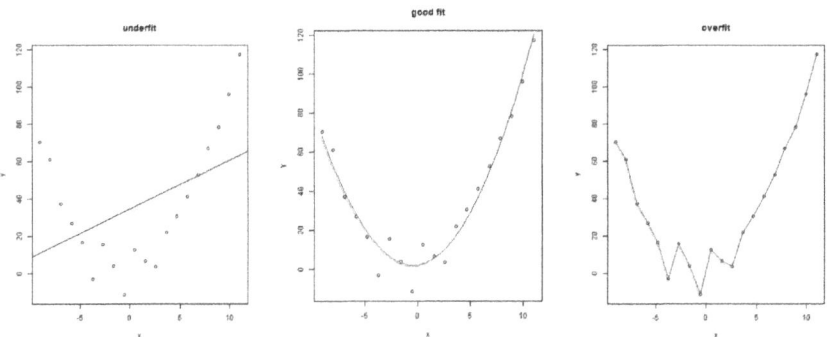

Fig. 5.7 A well-fitting model (centre) in between an underfit model (left) and an overfit model (right). (*Notes* Author's elaboration, see R code file for details)

Fig. 5.7 centre, we have a non-linear regression line that fits the data fairly well. It picks up the main trends, although it does not perfectly predict the positions of all the datapoints. On the right of Fig. 5.7, we have the case of overfit. The theoretical model fits all the datapoints exactly. It seems too good to be true. Such a theoretical model might perform well on the data it is trained on, but it is unlikely to perform well when applied to new data.

The problem of **overfit** refers to the idea that a model that works well in your **historical training data**, might not fit well to new data that is collected later (referred to as **test data**). There is the crucial distinction between the **in-sample** predictive power which comes from fitting a model to training data, and the **out-of-sample** predictive power that comes from applying the model to new test data. With modern data science, there are powerful tools for fitting models to data, and overfitting is a real problem. Given that these data-fitting techniques are so powerful, we do not care at all about the R^2 statistic (or model fit) that comes from fitting the model to the historical training data, the only thing that matters is the out-of-sample R^2 statistic that comes from applying the model to new test data. We don't care at all about the in-sample R^2 from the historical training data, because we can always increase the in-sample R^2 just by adding spurious variables and overfitting.

Overfitting is particularly problematic with **high-dimensional datasets**, for example if you have hundreds of explanatory variables (sometimes referred to as "**features**") in your regression model. In the classic case of OLS linear regression, the number of explanatory variables is much smaller than the number of observations. If the number of variables p is greater than or equal to the number of observations n, then OLS perfectly predicts the data. In the case of two observations, if you have an intercept term and a regression slope, then you can be 100% sure that you can fit a straight line to connect these two observations. The resulting linear model would have perfect predictive power on the historical training data, but will perform badly when applied to new data. A regression model trained on historical data may even have a negative R^2 statistic when fitted to new data (Taddy et al. 2023).

Recall the equation for the R^2 statistic:

$$R^2 = 1 - \frac{SS_R}{SS_T}$$

If the predicted values from our overfitted model are further away from the actual values than the predictions from the null model (where our prediction of y is simply the average of y, i.e. $\bar{y}$), then we are in the case of a negative R^2 statistic, and the overfitted model performs worse than having no model at all.

5.6.2 An Introduction to Lasso Regression

Lasso is a big data technique that is designed precisely to deal with the risk of overfitting that comes from powerful data-fitting techniques and high-dimensional data.

Let's first consider the well-known OLS linear regression model:

$$y_i = \beta_0 + \beta_1 x_{1i} + \beta_2 x_{2i} + \cdots + \beta_p x_{pi} + \varepsilon_i$$

We can rewrite this in vector notation:

$$y_i = x'\beta + \varepsilon_i$$

With OLS, the idea is to fit a regression line that minimizes the Residual Sum of Squares (SS_R). Put differently, OLS minimizes the following:

$$\sum_{i=1}^{n}(y_i - x'_i \beta)^2$$

With Lasso, the difference is that you add on a penalty term, that puts a price on model complexity:

$$\lambda \sum_{i=1}^{n}(y_i - x'_i \beta)^2 + \lambda \sum_{j=1}^{p}|\beta_j|$$

The penalty term has the effect of shrinking the regression coefficients to be as close to zero as possible.

If you force the estimated coefficients $\hat{\beta}_j$, to be closer to zero, then the predicted $\hat{y}$ values ($\hat{y}$) will be closer to $\bar{y}$, the average value of y. With Lasso, the estimated regression coefficients $\hat{\beta}_j$ have to justify why they should take non-zero values.

We now compare OLS and Lasso in the case of 6 explanatory variables. 6 variables is a very small number of variables for discussions of Lasso. Overfitting is not so important when you only have 6 explanatory variables, but it is more important when you have 600. Nevertheless, in this example we have just 6 explanatory variables.

With OLS, you have many coefficients that are nearly zero, but it is very unlikely that the coefficients are exactly zero. That means you have non-zero coefficients and a long list of explanatory variables.

$$y_i = \beta_0 + \beta_1 x_{1i} + \beta_2 x_{2i} + \beta_3 x_{3i} + \beta_4 x_{4i} + \beta_5 x_{5i} + \beta_6 x_{6i} + \varepsilon_i$$

With Lasso, the penalty term shrinks unimportant coefficients to zero, and then simplifies the model. Here, we have coefficients on x_2, x_3 and x_5 that are shrunk to exactly zero. This means that we can remove these three terms (shown here in faint ink) from our Lasso regression equation, which simplifies our model.

$$y_i = \beta_0 + \beta_1 x_{1i} + \beta_2 x_{2i} + \beta_3 x_{3i} + \beta_4 x_{4i} + \beta_5 x_{5i} + \beta_6 x_{6i} + \varepsilon_i$$

Lasso is useful because it is a meaningful way of getting rid of relatively unimportant predictor variables. An alternative approach could be manually checking each of the explanatory variables and deciding whether to keep it in or throw it out of the regression model, but this hand-picking of variables would take a lot of time and is less appropriate for big data contexts.

Figure 5.8 shows how the Lasso algorithm shrinks the regression coefficients towards zero. Figure 5.8 (left) corresponds to the case of an OLS regression where the coefficients are normally distributed around a mean of zero. Many coefficients are close to zero, but a negligible fraction would be exactly 0.000000. Irrelevant variables lead to unnecessary complexity in the regression model, and lead to the problem of overfitting, so we would prefer to drop them. We can remove irrelevant variables by setting their coefficients to be precisely zero. OLS linear regression would give these irrelevant variables a small non-zero coefficient, but OLS is extremely unlikely to give coefficient estimates that are exactly zero. Lasso shrinks the coefficients towards zero, and each coefficient is shrunk towards zero by a roughly constant amount (James et al. 2021, pp. 247–249). Lasso shrinks some coefficients to be exactly zero, which explains why we can also think of Lasso as a technique for variable selection, because an explanatory variable with coefficient zero is effectively dropped from the regression model. Figure 5.8 (right) shows a histogram of the corresponding coefficient estimates for Lasso, where (compared to the unconstrained OLS model) the Lasso coefficients are shrunk towards zero by an approximately constant amount, which leads to a mass point of coefficients at a value of exactly zero: and then these coefficients are dropped from the Lasso model. Note, however, that Lasso does not try to shrink the value of the regression intercept term.

5.6 Lasso Regression

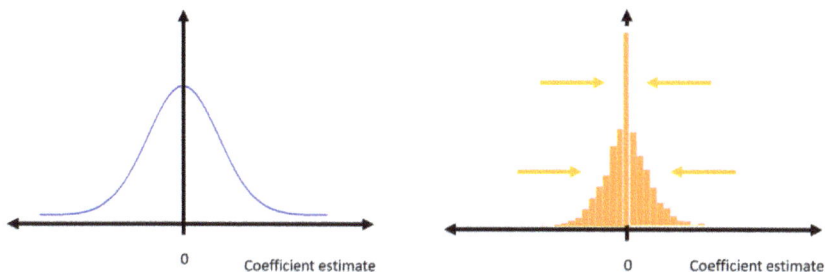

Fig. 5.8 A rough intuition of how Lasso shrinks regression coefficients to zero. Left: unconstrained OLS model. An OLS regression might yield coefficient estimates that are normally distributed around zero. Right: corresponding histogram of regression estimates from Lasso. Arrows indicate Lasso shrinkage. Lasso shrinks all the coefficients towards zero, by an approximately equal amount, leading to a mass point of coefficients at a value of precisely zero. (*Source* Author's elaboration)

5.6.3 R Example: Lasso Regression

The overall Lasso procedure boils down to two steps. First, we get 100 model estimates corresponding to 100 different values of the penalty term lambda. Then, in the second stage, we select the best model from these 100 candidate models, using a criterion such as the **AICc**, the corrected version of the AIC (Akaike Information Criterion). A non-trivial part of the work is getting the data into the right format for Lasso estimation in R, using the gamlr package.

Care is needed to prepare the data for Lasso estimation with the gamlr command. For continuous variables (referred to as numeric variables), you need a column for each variable. For categorical variables (or factor variables), you need a column for each level of the variable. If the variable is binary, it has two levels, and so will require two columns. gamlr requires a full set of levels for the variables. Do not drop the baseline reference case! When penalizing coefficients, the factor reference levels now matter! For example, with OLS regression, if we wanted to control for the four seasons, we would set one season (e.g. Spring) as the omitted baseline reference case, and include three dummies (one for Summer, one for Autumn, and one for Winter). With Lasso, we would not have three dummies, but four (or potentially five) dummies: one each for Spring, Summer, Autumn, and Winter, and perhaps another dummy to control for the group of missing values (Taddy et al., 2023), if there were any missing values for the seasons variable. The Lasso algorithm will then decide which seasonal dummies to drop. If we were to fix the omitted baseline case ourselves (e.g. omitting the Spring dummy), then this might affect the coefficient values for the other seasonal dummies. Instead of choosing the factor reference levels at the start, we therefore let the right choice of factor reference level be decided by the Lasso algorithm. The Lasso algorithm shrinks the coefficients toward a shared mean, with only significantly distinct effects getting nonzero coefficients.

We need to delete the intercept term (which is a column of ones) as we prepare the data for `gamlr`. We also need to convert the data to a sparse matrix format, which helps to reduce storage needs and increase efficiency, thereby requiring less memory and allowing for faster optimizations.

For Lasso to work well, the variables should be on comparable scales. In other words, your model fit should not change if a measurement is in milimetres, centimetres, or metres. Fortunately, `gamlr` automatically standardizes the variables. Each regression coefficient beta's penalty is measured in terms of 1 standard deviation change in the explanatory variable x.

Our data are free to download from the website of the European Commission's EU Industrial R&D Investment Scoreboard.[2] This is data on the world's largest R&D investing companies. We start with 21 variables, and 2500 firms in the data. The top-ranking R&D investors in this dataset are Alphabet, Meta, Microsoft, Apple, and Huawei.

Then we do some data cleaning and preparation to get the data ready for Lasso. First, we tidy up the column names, to make it easier to understand the data columns. Second, we convert some variables to become factor variables (that is, categorical variables). Third, we take logarithms (or the IHS, Inverse Hyperbolic Sine transformation) for some variables. Fourth, we remove unnecessary variables. Fifth, we decide what to do with cases of missing values.

Now we have to build the model matrix for `gamlr`, where `gamlr` is the command that we will use for Lasso. In our case, the dependent variable, that we are trying to explain, is the R&D intensity of firms. We create a separate vector (called `ysb`) for the response variable y. Then, we create a matrix of features (or explanatory variables) in sparse matrix format, and remove the response variable y (referred to in the R code as `-ycol`) and also remove the intercept term by adding "−1" at the end.

Figure 5.9 shows a lasso path plot, which is an interesting piece of the Lasso output in R. On the horizontal axis, at the bottom of the graph, we see "log lambda", which represents the values of the Lasso penalty term λ. On the vertical axis: we have the sign and magnitude of the (standardized) coefficient. Along the top of the graph, we have the number of non-zero coefficients, starting out at 98 on the left, and ending at 1 on the right. At the extreme left, the value of the penalty term is so low, that it has negligible effect. At the extreme left, we have about 98 variables in the regression model, and the regression model is similar to a standard OLS regression that includes all of the explanatory variables. At the extreme right, the value of the penalty term lambda is very large, and this means that most coefficients are shrunk down to zero. At the very extreme, there is only one variable that remains in the model. Each vertical slice of this graph corresponds to a regression model. This graph actually plots the output of 100 regression models.

---

[2] See here: https://iri.jrc.ec.europa.eu/scoreboard/2024-eu-industrial-rd-investment-scoreboard#field_data [last accessed 31 July 2025].

5.6 Lasso Regression

The best regression model, as selected by the AICc, is the model represented by the dashed vertical line in the middle.

We can then examine the gamlr output, shown in Fig. 5.10. There are 100 different regression models in the matrix beta, which correspond to the 100 different penalty weights (or segments) of lambda. The coefficient matrix beta has dimensions 101 times 100. There are 101 variables, and 100 different regression models. Because of the penalty term lambda, most of the models have far fewer than 101 variables, because most of the explanatory variables will be shrunk towards zero and then excluded from the regression model.

The best fitting model is the model at segment number 54. We can see that, in this model, lots of coefficients have been shrunk down to zero. If we drop the intercept term, we can see that there are 20 nonzero coefficients. The 5 most negative coefficients are country dummies for Indonesia and for Switzerland, the logarithm of employment, a country dummy for Japan, and a dummy for the Automobiles and Parts sector. The 5 most positive coefficients are a country dummy for China, some sector dummies (for the sectors Technology Hardware & Equipment, Pharmaceuticals and Biotechnology, and Software and Computer Services), and the logarithm of sales.

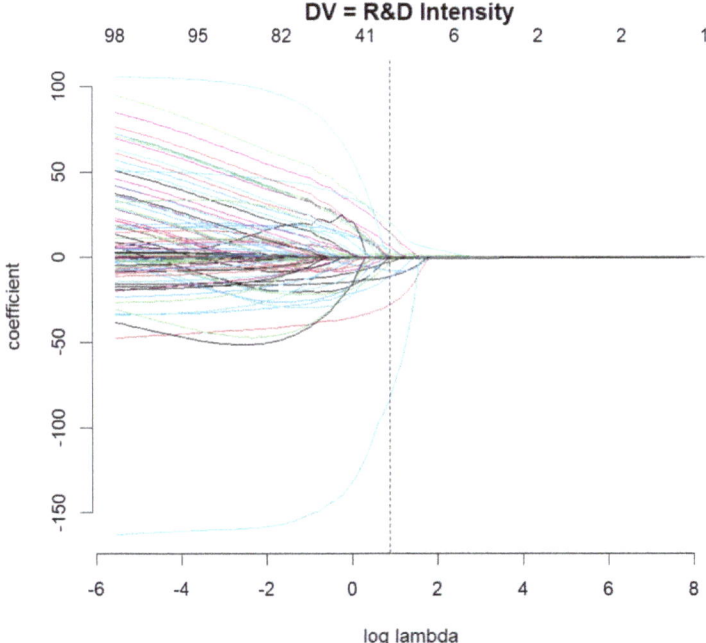

Fig. 5.9 Lasso path plot, for Scoreboard data. (*Source* Author's elaboration, see R code for details)

```
> # examine the gamlr object
> fitsb

gaussian gamlr with 101 inputs and 100 segments.

> # summary(fitsb)
> names(fitsb)
 [1] "lambda"   "gamma"    "nobs"    "family"  "alpha"   "beta"    "df"      "deviance" "iter"    "free"
[11] "call"
> dim(fitsb$beta)
[1] 101 100
```

Fig. 5.10 gamlr output, for Lasso regression. (*Source* Author's elaboration, see R code for details)

```
>
> # check coefficients for specific variables
> bsb[c("profbty", "sales_gr", "log_mcap", "empl_gr")]
     profbty     sales_gr     log_mcap      empl_gr
-0.746535285 -0.005991239  0.000000000  0.000000000
>
```

Fig. 5.11 Lasso output: coefficients. (*Source* Author's elaboration, see R code for details)

Further insights could be obtained by checking the coefficients for specific variables (see Fig. 5.11). Here the coefficient for profitability is -0.747, while the coefficient for sales growth is -0.00599. Both of these variables seem to be negatively associated with the dependent variable, which is R&D intensity. The coefficients for the two other variables, logarithm of market capitalization, and employment growth, are shrunk to zero, and are therefore dropped from the regression model.

And finally, another thing we can do is to use our Lasso-selected regression model to focus on the predicted values of the outcome (which is R&D intensity) for specific firms. Figure 5.12 focuses on the two largest R&D investors, Alphabet and Meta. We can see that the actual R&D intensities are 14.0% for Alphabet, and 28.8% for Meta. Then we can compare these to the predicted R&D intensities, which are 3.53% for Alphabet, and 18.59% for Meta. The difference between the actual R&D intensities, and the predicted R&D intensities, suggest that Alphabet and Meta have higher R&D intensities that we might think, based on their characteristics. The difference between the actual R&D intensities, and the predicted R&D intensities, could also suggest that our model for predicting a firm's R&D intensity is not always so accurate.

5.7 Alternative Regression Models

In this section, we saw how OLS works best with dependent variables that are continuous, whereas logistic regression works best for binary dependent variables.

There are many other types of regression models (Table 5.2), and the choice of regression model depends on the characteristics of the dependent variable. The

```
> 
> # Focus on the 2 largest R&D investors (i.e. Alphabet & Meta)
> sb_raw[ c(1, 2) , c(2, 7, 11) ]
# A tibble: 2 x 3
  Company  `R&D (€ million)`  `R&D intensity (%)`
  <chr>              <dbl>              <dbl>
1 ALPHABET          37034.               14.0
2 META              31520.               28.8
> # Actual R&D intensity = 14.0% for Alphabet, 28.8% for Meta
> 
> # The predict command uses the AICc by default
> ( yhat <- predict(fitsb, xsb[ c(1, 2) ,]) )
2 x 1 Matrix of class "dgeMatrix"
     seg54
1  3.525496
2 18.592296
> drop(yhat) # use "drop" to remove sparse formatting
        1         2
 3.525496 18.592296
> # Predicted R&D intensity = 3.53% for Alphabet, 18.59% for Meta
> 
```

Fig. 5.12 Lasso analysis: predicted values. (*Source* Author's elaboration, see R code for details)

important thing is the dependent variable; the characteristics of the explanatory variables (continuous, binary, percentages, integer numbers, etc.) are not important for choosing the regression model.

Table 5.2 Alternative regression models

Dependent variable	How many explanatory variables	Type of explanatory variable	Technique
Continuous	One	Continuous	Correlation (Pearson, Spearman)
Continuous	One or more	Continuous or categorical	OLS linear regression
Binary	One or more	Continuous or categorical	Logistic regression
3 + categories	One or more	Continuous or categorical	Multinomial logistic regression
Integer numbers	One or more	Continuous or categorical	Count data regression
Continuous but with a lower bound	One or more	Continuous or categorical	Tobit regression
Percentages	One or more	Continuous or categorical	Fractional dependent variable (Papke & Wooldridge, 1996)

Source Author's elaboration

5.8 R Example: OLS Regression on Salary Data

This example uses (fictional) data to help decide upon the remuneration of new hires, using "**salary.xlsx**". Our task is to explain the salary of individual i in terms of some personal characteristics.

Model 1 looks at the relationship between Salary and Experience. The results table in Fig. 5.13 shows a positive coefficient of 0.245 (highly statistically significant) on the variable "Experience", and Fig. 5.14 suggests that the positive relationship between Experience and Salary is indeed linear.

It is possible that other factors, beyond experience, can help to explain salary. Our data allow us to go further and include additional explanatory variables in Models 2 and 3 (see Fig. 5.15).

Model 2 suggests that Salary is positively related to Experience and Performance. Further analysis shows that these coefficients are statistically significant from zero.

Knowing the OLS regression coefficients allows us to calculate predicted values. For example, according to Model 2, what is the predicted salary for person X who will soon start at the company, who has experience = 5 and Performance = 6? The answer is calculated as follows: Salary = 3.208 + (0.2214 × 5) + (0.2963 × 6) = 6.093

Consider now the case of Model 3, where we find out additional information on the same person X: that they are female. Their predicted value of Salary can

```
> # Model 1: only 1 explanatory variable
> summary(lm(Salary ~ Experience, data = salary))

Call:
lm(formula = Salary ~ Experience, data = salary)

Residuals:
    Min      1Q  Median      3Q     Max
-1.1654 -0.4123 -0.1596  0.2096  3.2181

Coefficients:
            Estimate Std. Error t value Pr(>|t|)
(Intercept)  4.79797    0.23516  20.403  < 2e-16 ***
Experience   0.24534    0.05642   4.349 5.61e-05 ***
---
Signif. codes:  0 '***' 0.001 '**' 0.01 '*' 0.05 '.' 0.1 ' ' 1

Residual standard error: 0.7755 on 58 degrees of freedom
Multiple R-squared:  0.2459,    Adjusted R-squared:  0.2329
F-statistic: 18.91 on 1 and 58 DF,  p-value: 5.606e-05
```

Fig. 5.13 OLS regression, model 1. (*Source* Author's elaboration, see R code for details)

5.8 R Example: OLS Regression on Salary Data

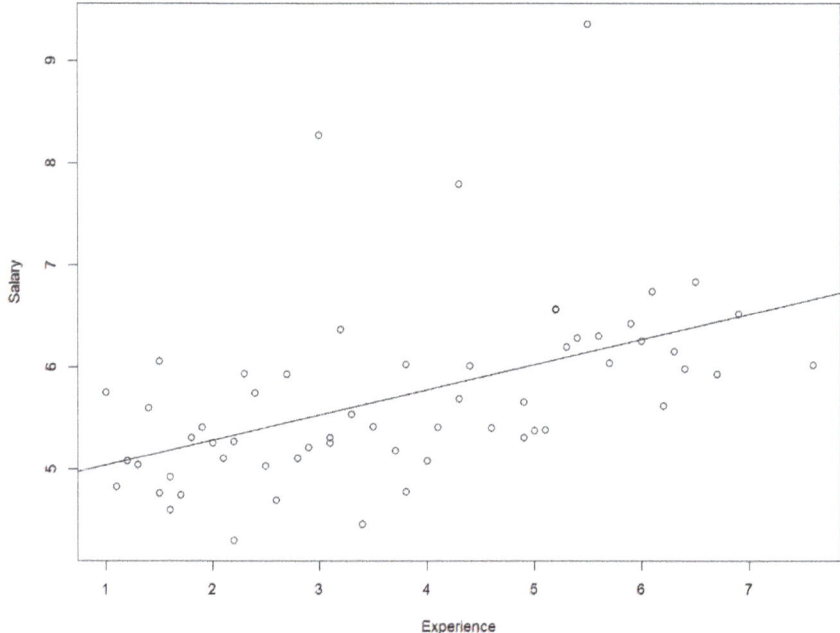

Fig. 5.14 Scatterplot of the relationship between Salary and Experience, with an OLS best-fit line overlaid. (*Source* Author's elaboration, see R code for details)

```
> 
> # Model 2: only 2 explanatory variables
> coefficients(lm(Salary ~ Experience + Performance, data = salary))
(Intercept)  Experience Performance
  3.2079416   0.2213960   0.2962715
> 
> # Model 3: all 3 explanatory variables
> coefficients(lm(Salary ~ Experience + Performance + Female, data = salary))
(Intercept)  Experience Performance      Female
  3.3184196   0.2206087   0.3031351  -0.3254147
> 
```

Fig. 5.15 OLS regression output for Models 2 and 3. (*Source* Author's elaboration, see R code for details)

be calculated in a similar way: Salary $= 3.318 + (0.2206 \times 5) + (0.3031 \times 6) + (-0.325 \times 1) = 5.9146$

Model 3 undeniably has a better fit to the data than Model 2, because the Adjusted R^2 statistic is higher (0.9591 vs 0.925), and all three explanatory variables in Model 3 are highly statistically significant. This leads us to the following question:

Do You Recommend Using This Predicted Salary from Model 3 to Inform Decision-Making?

The answer is: certainly not! The difference between Model 3 and Model 2 amounts to something like this: after taking into account your experience and performance, we decided to pay you a lower salary for no other reason than because of your gender.

This example has a serious message: just following the data can get you into trouble, because you can end up making the same biased decisions that occurred in the unglorious past. It is illegal to pay someone less according to gender. Saying that you were just following the data is not a satisfactory explanation. If a manager says that the decision to pay less to a female was driven by the statistically significant results from Model 3, such a justification will be of little use in a court of law.

Most of my students make this mistake each year. Business data analysis should be approached from two different and unrelated angles: first, whether the results are statistically significant and quantitatively important, and second, whether the results may directly or indirectly be unethical. Model 3 is better than Model 2 on purely statistical grounds, but in the world of business data analytics we cannot make our decisions on purely statistical grounds. The next chapter continues this discussion by looking into the ethics of data science and AI.

Further Reading

Imai and Williams (2022) focus on regression models with R examples, from an econometrics perspective. Taddy et al. (2023) focus on regression models with R examples, from a more data science perspective.

References

Angrist, J.D. and Pischke, J.-S. (2008). Mostly harmless econometrics: An empiricist's companion. Princeton University Press.
Greenland, S., Senn, S. J., Rothman, K. J., Carlin, J. B., Poole, C., Goodman, S. N., & Altman, D. G. (2016). Statistical tests, P values, confidence intervals, and power: a guide to misinterpretations. European Journal of Epidemiology, 31(4), 337–350.
Imai, K., & Williams, N. W. (2022). Quantitative Social Science: An Introduction in Tidyverse. Princeton University Press.
Provost, F., & Fawcett, T. (2013). Data Science for Business: What you need to know about data mining and data-analytic thinking. O'Reilly Media, Inc.
Taddy M., Hendrix L., Harding M.C. (2023). Modern Business Analytics. Practical Data Science for Decision Making. McGraw Hill, New York, NY.
Tibshirani, R. (1996). Regression shrinkage and selection via the lasso. Journal of the Royal Statistical Society. Series B (Methodological), 267–288.
Wickham H., Çetinkaya-Rundel M., Grolemund G. (2023). R for data science: import, tidy, transform, visualize, and model data (Second Edition). O'Reilly Media, Inc. Free to read online: https://r4ds.hadley.nz/

Ethics of Data Science and AI

We need to talk about ethics.

The digital storage of information has the potential to make information easier to gather, store, and access. As such, digitalization can potentially lead to greater transparency and greater accountability. This could potentially lead to a reduction in the frequency of unethical behaviour.

Data can lead to new types of ethical problems, however. Machine learning algorithms and AI can lead to ethical problems that can be very expensive for firms and organizations, leading to large fines, and reputational damage with customers and stakeholders. Given that ML and AI are designed to operate at scale, the risks are always large in scope (Blackman 2022, p. 5). As such, it is critically important to set things up properly with regards to ethics. This involves setting up corporate structures to minimize the risk of ethical problems, by consulting with ethics experts and legal experts.

Ethical risks come from many different use cases of AI, such as facial recognition, lip-reading AI, autonomous vehicles. The goal of AI ethics is not so much to do good, as to avoid the bad. AI ethics does not seek to maximize the positive social impact, but instead seeks to mitigate ethical risks. It is not so much about being "passionate" about stuffing AI ethics everywhere throughout your organization, but doing what it takes to avoid risks, that may never actually arise (but if they do, they could be absolutely catastrophic). The task is not to aggressively drive AI ethics into your operations, but instead to nestle AI ethics into operations to avoid the many possible ethical problems that could potentially appear

---

Supplementary Information The online version contains supplementary material available at https://doi.org/10.1007/978-981-95-2433-4_6.

(Blackman 2022). The challenge for firms is to develop an ethical strategy that is coherent and compatible with their business model, to articulate their ethical values clearly, and to explain how they fulfil these ethical values. Coherence is an important theme. For example, firms do not need to promise full transparency and explainability of their algorithmic decisions, especially if this is not especially useful for customers. Firms should not promise full customer privacy if their business model requires collecting and selling customer data.

6.1 Good News: Data can Elucidate how Decisions are Made

Technology is not intrinsically good or bad, what matters is how we use it.

On the bright side, data can help to spot unethical behavior that previously might have been unnoticed. For example, data allowed AirBnB to discover that distinctively-named African-Americans were less likely to get a successful booking, which allowed them to introduce initiatives to target this problem (Haenlein et al. 2022).

Decisions can be considered to be the outcome of two inputs: **prediction** and **judgment** (Agrawal et al. 2022, p. 155, p. 235). Data and AI help to generate predictions in a relatively low-cost way. Then, these predictions can be combined with value judgments from humans, regarding the comparative benefits of the different outcomes. Prediction generates probabilities, while judgment generates the payoffs for each outcome (i.e. the values we attach to each outcome). Data and AI help to separate the prediction from the judgment

These ideas are shown in Fig. 6.1. On the left, there is the usual case of an unhurried driver who seeks to arrive safely at the destination. We attach a utility of 20 to the outcome of arriving at the destination, and the probability of arriving safely is 99%. There is a 1% chance that we have a nasty accident, which is an outcome that gives us a utility of -1000. The overall payoff is $(20 \times 0.99) + (-1000 \times 0.01) = 19.8 - 10 = 9.8$.

On the right, the driver puts more emphasis on the importance of arriving faster, for example to arrive in time for an important event. This driver chooses a different position on the tradeoff between the benefits of arriving faster, and the danger of taking more risks. In this scenario, the driver now accepts the higher chances of having an accident (10%, compared to 1% previously) because the payoff to arriving faster is large (now at 200). The driver has now changed their risk-taking preferences and accepts a higher probability of having an accident, which is not only dangerous for them, but also dangerous for other road users.

This example has highlighted a problem that was not visible before. It is unethical to expose other road users to this higher level of risk, simply because our preferences have changed and we want to arrive faster. Society would probably not agree that we can take more risks with other people's lives just because we are in a hurry. These probabilities associated with the outcomes (i.e. how the probability of an accident was deliberately raised from 1% on the left, to 10% on the

6.2 AI Ethics

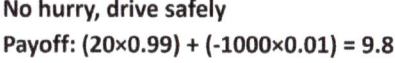
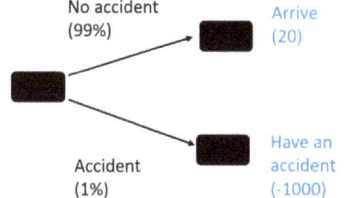
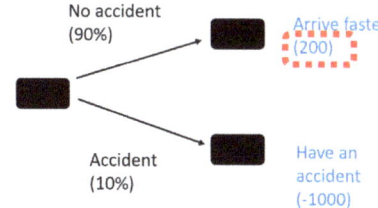

Fig. 6.1 Prediction, Judgment, and Decisions: 2 regimes with different payoffs. (*Source* Our elaboration, drawing on ideas in Agrawal et al. [2022])

right) are not measurable if the decision is made inside a human's mind, but could be visible if the preferences for driving fast, and varying levels of risk tolerance, are coded into the decision rules of an AI driver.

This example shows that, in normal times, human drivers may drive slowly to be extra sure to avoid accidents. But when they are in a hurry, human drivers may be more willing to take risks, and prioritize speed and accept higher chances of an accident. Consider the case of a self-driving car that receives visual signals with uncertainty. What was the probability that that obscure object that emerged from behind the school bus was a small child? Is the probability 0.1%, or 0.000001%? What would be the acceptable threshold? There are tradeoffs, that we might not think about until we examine the decision under uncertainty made by the AI. Interestingly, for humans, we don't separate the decision into prediction and judgment. But for autonomous vehicles, it is easier to distinguish between the prediction and the judgment. AI decisions can be scrutable in ways that human decisions are not. This could be an advantage of AI.

6.2 AI Ethics

The driving example in the previous Sect. 6.1 showed how AI can help shed light in areas that were previously not explicitly scrutable, such as preferences for risky driving and the varying notions of acceptable amounts of risk. In the driving example, AI led to better outcomes because it led to the collection of new data on matters that previously had not been studied.

However, AI can also lead to ethical problems. There are three main topics for AI ethics, which are biased AI, black-box algorithms, and violations of privacy (Blackman 2022, p. 11). These are discussed in the following subsections.

6.2.1 AI Bias Due to the Training Data

There are many examples of shocking injustices that have been carried out when AI and Machine Learning algorithms, trained on past data, make biased predictions. For example, a criminal justice algorithm in Florida misclassified African-American defendants as "high-risk" at about twice the rate that it misclassified white defendants (Manyika et al. 2019). During the 2020 Coronavirus pandemic, which led to the cancellation of school final exams in the UK, exam grades were replaced by AI-predicted grades, which resulted in a bias in the form of systematically lower predicted grades for students from poorer backgrounds.[1]

Figure 6.2 highlights how this bias emerges. To begin with, at the top, biased decisions, are made that will include decisions based on racist, sexist and homophobic stereotypes. These past decisions are recorded and stored in datasets. These datasets, in turn, form the basis for data analysis, the formation of decision-making models, and predictions and recommendations. The fourth stage is where these biased predictions and recommendations, formed from the biased datasets, are blindly followed and implemented by people to complete the circle, leading to the next generation of prejudiced decisions. To avoid the perpetuation of this cycle of biased behaviour, biased data, and biased recommendations, the loop needs to be cut somewhere. Probably the best way to do this is to raise awareness of the problems of biased data, learn from problematic cases in the past, and ensure that current decision-making is based not only on data analysis, but also critical thinking and familiarity with the ethical aspects of business decisions. The challenge is that we should not merely seek to replicate the world as it is today (with its many biases and prejudices), but to create a fairer world by correcting for previous biases.

Removing discrimination from models trained in biased data is notoriously difficult. Removing obvious sources of bias does not solve the problem, because it often leads to situations where bias takes more indirect and unexpected ways of continuing to influence the AI-based recommendations, as nicely summarized by Fourcade and Healy (2024, p. 243): "Group-level differences that the law kicks out of the door come back in through the window." Amazon spent considerable resources developing a hiring algorithm, which ended up being discontinued because it kept displaying bias in terms of preferences for male applicants. Even variables as subtle as the choice of words (such as "executed" or "captured" to describe actions in business situations) were indirectly involved in systematically preferring male applicants over females (Manyika et al. 2019). After a few years, Amazon abandoned this AI tool because it could not find a way to prevent it from discriminating against women (Blackman 2022, p. 4). If Amazon engineers could not solve this problem, clearly it is a tricky issue.

---

[1] https://www.forbes.com/sites/charlestowersclark/2020/08/25/uk-exam-results-U-turn-algorithms-alone-cant-solve-complex-human-problems/?sh=3f536f128107 [last accessed 23 June 2025].

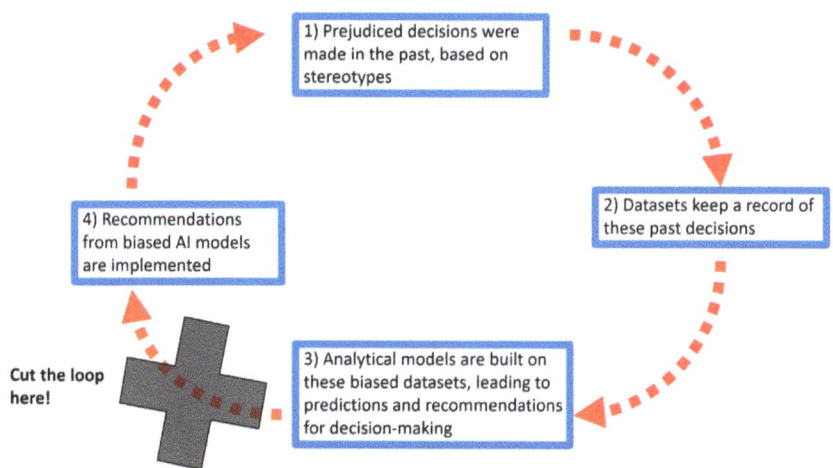

Fig. 6.2 How bias enters the AI loop, and how to cut the loop. (*Source* Author's elaboration)

There are various types of problems that emerge due to AI bias. Face recognition technologies have repeatedly been observed to have less precision and higher error rates for minorities (and in particular minority women). Such is the case of the "racist" soap dispenser that only squirts soap into the hands of white individuals, but that often fails to be activated by the hands of black individuals. Note that the problems of facial recognition software being biased against some racial minorities is due to the problem of insufficient observations in the training data (Blackman 2022). This is a different problem from the bias that affects AI hiring software, where the problems of bias are not because of insufficient data, but because the data contains ingrained bias, given that the data is based on prejudiced decisions made in the past. More data will help solve the problem of racist soap dispensers, but more data will not help solve the problem of racist hiring practices. Choosing appropriate strategies for mitigating AI bias requires expertise and a case-by-case approach, because what works in one context (collect more data to address bias due to some groups being under-represented in the training data) will not help in another context.

6.2.2 Black Box Algorithms and Explainability

Transparency is a virtue. It is generally preferable if we can explain the reasons behind our decisions, particularly if there is a lot at stake. As such, there has been a lot of frustration against AI decisions that seem to be unfounded or unjust, and there have been many calls for improving the explainability of machine learning and AI algorithms.

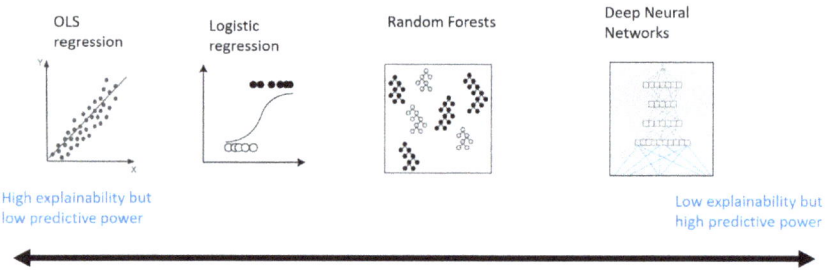

Fig. 6.3 A trade-off between the predictive power of techniques, and their explainability. (*Source* Author's elaboration)

Explainability is generally a desirable characteristic of machine learning and AI techniques, but explainability is not an end goal in itself. Explainability could be a strategy that a firm uses to protect its reputation and to build trust with users, but explainability is not something that should be maximized.

Explainability could potentially be in conflict with your business model, because higher explainability can come at the cost of less accurate predictions. Figure 6.3 suggests that there can be a trade-off between the predictive power of techniques, and their explainability. On the left, OLS linear regression is based on the idea of fitting a straight line through a cloud of datapoints, and has a relatively high explainability, but the predictive power is often lower than other techniques that automatically detect interactions and nonlinear patterns, such as Random Forests and Deep Neural Networks. Logistic regression is sometimes preferred to OLS regression in the case of binary dependent variables, but its explainability is less straightforward.

Explainability could also be quite unnecessary for some mundane activities, such as the programming of deliveries to suppliers based on the weather, local traffic conditions, and available space in trucks.

Explainability also comes at a cost, if it takes time and resources to articulate and communicate all the details behind the decision making, and if rivals may learn about the details of your business processes.

As such, machine learning and AI outputs do not always need to be fully explainable.

If explainability is paramount, you might need to use techniques that offer better explainability at the cost of worse performance in terms of predictive power. In other cases, however, accuracy matters more than explainability. Few would be willing to compromise on the accuracy of cancer prediction technology, for example, if higher explainability comes at the cost of lower accuracy (Blackman 2022).

Finally, another issue is that the appropriate explanation could take many forms. If a mortgage application is unsuccessful, should you explain which variables are used to make the decision? Or explain which technique was used to convert inputs into the output decision? Or explain why you took a particular number for the cutoff threshold to distinguish between successful and unsuccessful applicants? Perhaps more useful would be to offer a counterfactual explanation based on actionable steps to take in order to be successful next time (Kearns and Roth 2019), such as "stay in your current job for another 6 months", or "increase your initial deposit by 5%."

6.2.3 Privacy

AI technologies have many benefits but are also raising ethical concerns about privacy. For example, AI can be used for extreme employee surveillance, with software that can scrutinize staff behavior on a minute-to-minute basis, including information on who interacts with whom (whether by email or in-person interactions), and who accesses and edits which files (Haenlein et al. 2022). AI has also been a game-changer for marketing activities, allowing a more detailed analysis of customer information, more accurate predictions of individual-level willingness-to-pay, and raising concerns about the ethics of using powerful AI techniques to trigger impulse buying by vulnerable groups in product categories such as gambling, tobacco, and unhealthy food (Haenlein et al. 2022).

The "five Ps" approach (Provenance, Purpose, Protection, Privacy, and Preparation; see Segalla and Rouziès 2023) is a useful framework for thinking about AI and ethical data handling.

- Provenance refers to the source of the data, and questions regarding how it was acquired and whether appropriate consent was obtained
- Purpose relates to the ethical problems of taking data collected for one task and repurposing and reusing it in another context
- Protection focuses on how securely the data is being protected, and the conditions (who and when) regarding its destruction
- Privacy refers to whether (and how) the data will be anonymized, and which individuals will be able to access the data
- Preparation refers to aspects of how the data is processed, including data-cleaning, merging of datasets, preservation of anonymity, data quality checks (such as procedures for outliers and missing data), and so on

In each of these five areas, care is needed to ensure that the firm complies with the relevant regulations, takes necessary precautions against cyberattacks, and takes the appropriate behaviour regarding ethical risks.

One area of debate surrounds the possible tradeoffs between protecting privacy, on the one hand, and a growing AI industry's need for data, on the other. Do higher standards for data privacy come at the cost of lower growth of innovation in the digital economy and the AI industry (Haenlein et al. 2022)? According to the "Amazon flywheel" model (Baley and Veldkamp 2025), there exists a feedback loop whereby better data leads to better predictions, thus attracting more users, who generate more data, that can be used to train AI algorithms to get ever-better predictions, and so on, creating a virtuous cycle. Slowing down this flywheel effect could reduce the dynamism of the digital sector. However, there may be ways to avoid ethical problems linked to consumer privacy (e.g. keeping data safe from cyberattacks) that do not slow down the flywheel effect. Also, does the data generated by user activity belong to the user or to the platform owner? "Does the car owner own the data generated by her vehicle or can the manufacturer lay claim to it to (a) use the data, (b) resell the data, and (c) preclude the car owner from selling the data?" (Adner et al. 2019, p. 257). Transferring data rights from consumers to manufacturers could have negative effects on consumer well-being. One useful way of thinking about these ethical trade-offs is the "veil of ignorance" approach to designing AI rules, inspired by a thought experiment by philosopher John Rawls and applied by DeepMind (Abernethy et al. 2024, p. 64). According to this approach, rules for society are proposed by individuals who do not know what role they will have in that society, and don't know how the rules will affect them. Designing rules that are advantageous for one side (e.g. a data-greedy AI firm) would be regrettable if they were to end up being positioned on the other side (e.g. a vulnerable consumer). Decisions obtained from a "veil of ignorance" thought experiment tend to be less driven by self-interest and more sensitive to the situations of the most underprivileged members of society.

6.3 Competing Notions of Fairness

Firms that strive to be totally fair will eventually notice that it is an impossible task. Trying to please everyone with your strategy for an ethical approach to AI could prove futile.

For example, we probably agree that a person who commits a crime should be punished. But is it fair to give the same harsh punishments to young people compared to old people? Some would say that young people should be given a second chance, and punished less severely, especially if they started from a position of deprivation or if they lacked wisdom because of their youth. Another example could be that if there is an accurate prediction that a person will commit a crime, this knowledge should be acted upon. However, if this prediction emerged from analytics trained on data that relates to an individual's parents (and not the individual per se), should an individual face higher chances of punishment because of what their parents did (Blackman 2022, p. 77)?

The crux of the problem is that there is not just one definition of fairness, but many definitions, that may be in conflict with each other. Fairness can be defined in at least 21 different ways.[2] Table 6.1 gives an overview of some of the more

---

[2] See Narayanan, A. Translation Tutorial: 21 Definitions of Fairness and Their Politics. FAT* 2018. https://www.youtube.com/watch?v=wqamrPkF5kk [last accessed 25 June 2025].

common definitions of fairness. A first definition of fairness (i.e. statistical fairness) is that the analytical results should faithfully represent the statistical truths in the training data. However, a drawback of this indicator would be that it risks repeating existing biases such as sexism and racism, if for example white men are over-represented in the historical data as well as the model outputs. A second definition of fairness, procedural fairness, could be that all individuals have to go through the same process in order to get the same outputs. No allowances would be made for individuals with disabilities or those coming from disadvantaged backgrounds, which risks excluding certain groups. A third definition of fairness, distributive fairness, focuses on the fairness of the outcomes of the different groups. For example, individuals from minority groups could have preferential treatment in being promoted to leadership roles, if such minority groups were excluded from such roles in the past. Note that distributive fairness could conflict with statistical fairness and procedural fairness.

Other notions of fairness could arise if there are different priorities placed on the injustices of false positives and false negatives. Data scientists often refer to concepts such as false positives, false negatives, confusion matrices, and other terms such as sensitivity and specificity (see Table 6.2 for an example of a **confusion matrix**). A definition of fairness that focuses on minimizing the number of innocent individuals in prison (false positives) will lead to different decision rules than a definition of fairness that focuses on minimizing the number of guilty individuals who are still free to roam the streets (false negatives).

Finally, it would be nice if the error rates of false positives and false negatives were the same across groups (e.g. racial groups, gender groups). However, if groups differ in terms of their probability of e.g. a criminal defendant reoffending in future, then it could be mathematically impossible to have predictive parity (the same likelihood of recidivism among high-risk offenders regardless of group membership) at the same time as having the same error rates (in terms of false positives and false negatives) across groups (Chouldechova 2017).

In sum, there are many definitions of fairness that may even be in conflict with each other, and regulation does not always provide clear guidance on which definition of fairness matters most. The current situation is complex and problematic.[3]

6.4 Problems of Machines Lacking Accountability

Managers can work with AI in various ways, giving various amounts of autonomy to machine learning and AI in decision-making (HBR 2023). Table 6.3 outlines some different types of human–machine collaboration, ranging from **"Human in**

---

[3] See https://www.vox.com/future-perfect/22916602/ai-bias-fairness-tradeoffs-artificial-intelligence [last accessed 25th June 2025].

Table 6.1 Various notions of fairness, with definitions and examples

Notion of fairness	Definition	Example
Statistical fairness	The output should faithfully conform to the proportions in the real-world data	If most CEOs are male and most nurses are females, translation tools and generative AI should generally suggest that CEOs are male and that nurses are female
Procedural fairness	All individuals should face the same procedure	All individuals should go through the same procedures if they apply for a contract, no preferential shortcuts or fast-track routes
Distributive fairness	The algorithm is fair if it leads to fair outcomes	Some groups may end up cross-subsidizing others. There may also be preferential treatment for some groups, to repair for past injustices (e.g. higher promotion rates for female managers, to correct for a previous low share of females in the top management team)
Representational fairness	Each group should have the same outcomes	Three medals each for the men's and women's categories of the Olympic track events, even if some men without medals are faster than the gold-medal winning female
Minimize false negatives	It is not acceptable for guilty people to be misclassified as innocent	Focus on avoiding the situation where criminals walk free, even if this requires having a higher rate of putting innocent people in prison
Minimize false positives	It is not acceptable for innocent people to be misclassified as guilty	Focus on avoiding the situation where innocent people are punished, even if this requires having a higher rate of criminals on our streets
Same precision across groups	The precision should be the same across groups	When calculating the probability that a criminal defendant will reoffend, the error rate should not be higher for one racial group compared to another

Source Author's elaboration

6.4 Problems of Machines Lacking Accountability

Table 6.2 A confusion matrix, which is a contingency table of observed and predicted cases

		Actually observed		
		Positive	Negative	
Theoretical prediction	Positive	**True positive (80)**	**False positive (16)**	Positive Predicted value (PPV): 80/96 = 83.3%
	Negative	**False negative (3)**	**True negative (1)**	Negative predicted value (NPV): 1/4 = 25%
		True positive rate (TPR): 80/83 = 96.4%	True negative rate (TNR): 1/17 = 5.9%	Accuracy (ACC): 81/100 = 81%

Notes TPR is also known as Recall or Sensitivity. TNR is also known as Specificity. PPV is also known as Precision. Balanced Accuracy (BACC) here would be (TPR + TNR)/2 = (96.4% + 5.9%)/2 = 51.15%

Table 6.3 Varying degrees of human involvement in machine learning processes

Human in the loop (HITL)	The machine provides decision support or partial automation, known as "intelligence amplification" A human decision-maker is positioned in between the AI outputs and the decisions
Human in the loop for exceptions (HITLFE)	Humans handle the exceptions, and humans decide which cases are exceptions Codify a machine's level of confidence in its predictions, and then humans can check the cases where the machine had low confidence
Human on the loop (HOTL)	Humans review the decision outcomes and adjust the rules and parameters for decisions in future
Human out of the loop (HOOTL)	The machine makes all the decisions, the human intervenes only by setting new constraints and objectives Example: AI-based automated trading on financial markets

Source Based on information in HBR (2023)

the loop" (HITL) for the case where a human decision-maker mediates the AI outputs and the actual decisions; to the case of **"Human out of the loop" (HOOTL)**, for rare cases where the machine makes all the decisions, and the human rarely intervenes (for example, AI-based automated high-frequency trading on financial markets, and the autonomous ship called Mayflower[4]).

Nevertheless, machines themselves do not have moral responsibility. Machines do not have consciousness, free will, and the capability to form intentions, which

---

[4] See https://www.ibm.com/case-studies/mayflower [last accessed 31st July 2025].

are necessary if machines are to be considered as moral agents (Coeckelbergh 2020). There may be difficulties in matching an accident to the person responsible (e.g. whether the autonomous vehicle crashed because of the software team, or the component supplier, or the regulator, or the driver); nevertheless it would not be satisfactory to blame the machine while absolving the human side. "If you cannot see a human in the loop, you just need to look for a bigger loop" (Seaver 2018, p. 378).

AI and automation are attractive options for managers, because they hold the promise of more efficient decision-making. However, this efficiency is exactly what leads to a lack of accountability (Narayanan and Kapoor 2024, p. 12). The challenge for managers is to remember that humans are ultimately responsible for errors made by autonomous systems and algorithms, whether they be HITL or HOOTL systems.

The challenge for managers is also to consider that AI recommendations are only recommendations, and that they are frequently wrong, and that human collaborators frequently need to over-ride these AI recommendations. There is evidence, however, that humans give far more trust to AI systems and LLMs than they deserve (Narayanan and Kapoor 2024). All too often, where there is a discrepancy between what the machine recommends and what the manager thinks best, the manager often falls back on following the machine's advice (and then observes that the machine was mistaken). An example of this effect, known as "automation bias", would be the case of airline pilots who can see that the computer system is indicating a problem with the wrong engine, but they still shut down the wrong engine (Narayanan and Kapoor 2024). All too rarely do humans challenge the AI decision.

6.5 Expertise is Needed

AI ethics is indeed a difficult area. The ethical risks are potentially huge, potentially leading to fines and severe reputational damage. The ethical risks are also difficult to predict, and ethical risks (and the appropriate strategies for prevention of the worst outcomes) vary considerably from one industry context to another. The appropriate solutions to ethical problems are not always straightforward. Sometimes what seems like an ethical problem turns out to be a non-issue and can be ignored. For example, while explainability can be a virtue, explainability also has many costs that may not be justified, and so not all firms need to maximize explainability. Another example would be fairness: firms should not strive to be fair by all definitions of fairness—in fact this would be impossible. Instead, firms should focus on addressing their most important ethical risks, and be consistent.

Considerable expertise is required to choose the right metrics of fairness, and appropriate strategies for the mitigation of AI bias. As such, firms are advised to set up a committee such as an **AI Ethics Committee** (AIEC; Blackman 2022), or perhaps a corporate Institutional Review Board (IRB; Segalla and Rouziès 2023) that is similar to the ethics boards whose approval academics and medical researchers require in order to obtain access to research funding or permission to publish. Such an AIEC or IRB would be an extra bureaucratic step, for sure, although the benefits would include effectiveness in avoiding AI ethical risks. Such a committee would include about 4–7 members, and would be diverse in composition, including members representing data science, a business executive, a compliance specialist, a legal expert, and of course an ethicist (Blackman 2022; Segalla and Rouziès 2023). The committee would be guided by the firm's AI ethics statement, which sets out your ethical values and ethical nightmare scenarios, your strategic priorities, and the kinds of relationships you seek with your stakeholders. The committee would also be guided by its AI ethical case law, according to which committee members work through tricky ethical scenarios in advance, with a relatively detached and objective mindset where they are free from crises, pressures, and temptations to take shortcuts for short-term financial gain (Blackman 2022).

6.6 R Example: Logistic Regression on Loan Data

This example considers the case of whether loans were granted to individuals by a bank, using a fictional dataset ("**loan.xlsx**"). The model fitted to past data could potentially be used to make future decisions regarding which individuals should be granted a loan. The main dependent variable is a dummy variable for whether the loan was granted, and other variables include the individual's income, their credit rating, number of credit cards, age, education, home-owner status, whether the individual is a student, marital status, geographical region, average credit card balance, race (dummy variable equal to 1 for minority race), and whether the individual has had a previous loan.

We estimate these regression equations using logistic regression, and the results are in Table 6.4.

Model 1: $LoanGranted_i = f(\beta_0 + \beta_1 \, Income_i + \beta_2 \, CreditRating_i + \beta_3 \, PrevLoan_i)$

Model 2: $LoanGranted_i = f(\beta_0 + \beta_1 \, Income_i + \beta_2 \, CreditRating_i + \beta_3 \, PrevLoan_i + \beta_4 \, Race_i)$

Model 1 finds that some of the variables (Income and PrevLoan) are statistically significantly associated with whether the loan was granted. Individuals with higher income, and individuals who received a loan previously, are more likely to be granted the loan. This seems reasonable.

Model 2 expands upon Model 1 by adding a term for an individual's race. Model 2 has a better fit than Model 1, in terms of the AIC (the lower the AIC, the better the fit) and the Nagelkerke R^2 statistic. However, Model 2 could be problematic for ethical reasons, if an individual's race enters in to the calculations for predicting which individuals are granted a loan. One approach could be to prefer Model 1 over Model 2, because Model 1 does not explicitly include a variable for

Table 6.4 Logistic regression on the loan data: coefficients and z statistics

Dependent variable	Model 1 LoanGranted	Model 2 LoanGranted	Model 3 PrevLoan
Constant	−1.7321	−1.5864	−1.7007
	−7.83	−6.99	−7.243
Income	0.0165	0.0167	0.0154
	4.23	4.23	3.871
CreditRating	−0.00028	−0.00039	−0.00023
	−0.33	−0.45	−0.251
PrevLoan	0.9897	0.9455	
	5.45	5.18	
Race		−0.5887	−0.7255
		−2.59	−2.883
AIC	905.8	900.6	832.92
Nagelkerke R^2	0.1542	0.165	0.095

Notes Author's elaboration, see R code file for details

race, even though this data is available in the dataset. However, even Model 1 is likely to be affected by racial bias.

Model 3 takes a different dependent variable, and finds that race has a significantly negative association with having previously received a loan. Perhaps a person's race was used in the previous period to decide how to grant loans. If so, it would be problematic to use the outcome of the previous loan decision as an input into the current loan decision, because the biases in previous decisions will continue to exert an effect on current decisions. Further regressions (not shown here, see Models 4 and 5 in the R code file) show that race is also associated with lower income and lower credit ratings. As such, basing current decisions for granting loans on variables such as income and credit ratings is ethically problematic, unless efforts are made to correct for previous biases. When previous decisions were affected by bias, it is notoriously difficult to remove past biases from current decisions. Simply avoiding to include race as an explanatory variable in the decision to grant a loan is not sufficient, because bias can indirectly creep back in, through unexpected channels.

Further Reading

Kearns and Roth (2019) and Blackman (2022) are highly recommended books on the ethics of data science and AI.

References

Abernethy, J., Candelon, F., Evgeniou, T., Gupta, A., & Lostanlen, Y. (2024). Bring Human Values to AI. Harvard Business Review, 103(3–4), 59–68.

Adner, R., Puranam, P., & Zhu, F. (2019). What is different about digital strategy? From quantitative to qualitative change. Strategy Science, 4(4), 253–261.

Agrawal A., Gans J., Goldfarb A., (2022). Power and Prediction. Harvard Business Review Press, Boston, Massachusetts.

Baley, I., & Veldkamp, L. L. (2025). The Data Economy: Tools and Applications. Princeton University Press.

Blackman, R. (2022). Ethical Machines. HBR Press, Cambridge: MA, USA.

Chouldechova, A. (2017). Fair prediction with disparate impact: A study of bias in recidivism prediction instruments. arXiv e-prints, arXiv-1703.

Coeckelbergh, M. (2020). AI ethics. MIT Press. Cambridge, MA: USA.

Fourcade, M., & Healy, K. (2024). The Ordinal Society. Harvard University Press: Cambridge, MA, USA.

Haenlein, M., Huang, M. H., & Kaplan, A. (2022). Guest editorial: Business ethics in the era of artificial intelligence. Journal of Business Ethics, 178(4), 867–869.

HBR. (2023). HBR Guide to AI Basics for Managers. Harvard Business Review Press. Massachusetts: USA.

Kearns, M., & Roth, A. (2019). The ethical algorithm: The science of socially aware algorithm design. Oxford University Press.

Manyika, J., Silberg, J., & Presten, B. (2019). What do we do about the biases in AI. Harvard Business Review, 25.

Narayanan, A., & Kapoor, S. (2024). AI Snake Oil: What Artificial Intelligence Can Do, What It Can't, and How to Tell the Difference. Princeton University Press.

Seaver, N. (2018). What should an anthropology of algorithms do? Cultural Anthropology, 33(3), 375–385.

Segalla, M., & Rouziès, D. (2023). The Ethics of Managing People's Data. Harvard Business Review, 101(4), 86–94.

Working with Data 7

Data cleaning is a pain, and it takes a lot of time. Before you can start analyzing data, or before you can hand over the dataset to your Machine Learning engineers, there is a lot of preparatory work to be done. This preparatory work includes collecting the data, cleaning it, engaging in exploratory data analysis to get a better feel for the data, and checking for any suspicious elements that look like they could be errors.

This chapter discusses data quality, common problems of working with data, and data preprocessing. Then, an example in R revisits some of these themes.

7.1 Data Quality

Data analysis is largely about data cleaning. "It is typical to hear that 75–80% of an analyst's time is spent sourcing, cleaning, and preparing data for analysis." (Davenport 2014, p. 19). A lot of available data sources are of dubious quality and cannot be trusted. As such, data quality issues take up about 80% of data scientists' time (Kenett and Redman 2019). It is the one problem that they complain about the most. One can never be sure about whether all the errors have been found. Furthermore, these problems become larger in the age of AI: poor quality data can appear in the historical data that is used to train the statistical model, and appear again when that model is applied to new data. A prudent approach would be to not trust the data, until the data quality has been proven (Kenett and Redman 2019).

Supplementary Information The online version contains supplementary material available at https://doi.org/10.1007/978-981-95-2433-4_7.

Reliable data analytics require that the data be of a high quality. Kenett and Redman (2019) put forward three criteria for high quality data. First, the data should be "right": it should be factually correct and properly labelled. Second, the data should be "the right data" in the sense of being representative, unbiased, comprehensive, and relevant for the research questions and purposes of the analytics. Third, the data should be represented in the right way, in an appropriate format for being imported into the analytics environment. For example, people cannot read bar codes. Software packages cannot usually read image data in jpg format.[1]

Data quality issues are also present in the case of automated measurement done by Internet of Things (IoT) devices (Kenett and Redman 2019). Automated measurement may be free from human error, but there are still many sources of error that may arise, such as sand blocking the rotation of an anemometer (wind cone), or an electricity grid meter that suffers from intermittent failures.

An important first step is to ask if the dataset's quality is good enough for analysis (Kenett and Redman 2019). If the data was created in conformity with reliable data quality procedures, then it can probably be trusted. Otherwise, it requires a more detailed verification.

Checking a dataset's quality can proceed with an iterative process. To start, you could take a sample (of the 50–100 most recently created cases) and see if the data quality appears to be satisfactory. Obvious errors can be highlighted. If there are fewer than 5% of cases that have an error, then the data can be used with caution. If there are 20 errors, and 18 of them occur in the same variable, then the overall data quality could be improved by eliminating this variable. Obvious errors can be replaced with a missing value, such as "<NA>" in R. In some cases, errors can be corrected with their likely values, based on your contextual knowledge. In other cases, errors can be corrected through a deep study, checking each case. If the same errors tend to be repeated, automated techniques can be used to replace errors with their plausible values. Errors in data should be eliminated, and do not hesitate to classify datasets as uncertain and unreliable if need be. For many data science purposes, data quality does not have to be perfect for it to yield valuable new insights, but caution is needed to understand where the errors are, and to deal with errors, while being ready to abandon some datasets if the quality is too poor.

An often-heard phrase regarding data quality is "**Garbage In, Garbage Out**", with the acronym **GIGO**. Data sets in the wild differ considerably from datasets in data science tutorials. Unlike in the case of didactic datasets used in statistics courses, in the real world you cannot expect to receive a perfectly formed dataset. As such, the skill sets taught at university courses on applied statistics seem somewhat incomplete for industrial data science contexts. Poor quality data is a real problem and can affect everything your entire organization does. Much of the data used in analytics is also used in other divisions such as in basic operations and logistics. Errors in the address details may slow down your data science analyses,

---

[1] See however Taddy et al. (2023, Chap. 10) for an exception, where R imports jpg files for image classification via deep learning.

	A	B	C	D	E	F
1	ID	First name	Surname	Title	Town	Country
2	7754	Andrew	Smith	Mr	Leicester	United Kingdom
3	9352	Jane	Jonson	Ms	Loughborough	United Knigdom
4	9352	J	Jonson		Loughborough	UK
5	8655	Angela	Jensen		Quorn	United Kingdom
6	6032	Jessica	Aldershot	Mr	Leicester	UK
7	7735	Robert	Elgar		Loughborough	
8	3670	Christina	Robinson	Ms	Kegworth	United Knigdom

Fig. 7.1 Examples of data collection errors, including duplications (in orange), missing values (in green), and inconsistencies (in blue). (*Source* Author's elaboration)

but could also mean that a customer's package was not delivered on time, or not delivered at all. Clearly, sending merchandise to the wrong address is a waste of time and money. Research has shown that data quality issues typically cost an organization 20% of revenue (Redman 2017).

7.2 Common Problems of Working with Data

Recording errors are a fundamental issue in data collection. An illustrative example is from Segalla and Rouziès (2023), who used gender data from a large international company, as part of their research into the links between gender and the career benefits of training. Instead of finding a tidy binary variable for gender, there were 94 different values, that included the following: Woman, Female, F, f, Feale (typo; should be Female), Mujer (Spanish), Frau (German), and so on. Other variables were trickier still: "Employment start and end dates were especially problematic because of differing formats for dates" (Segalla and Rouziès, 2023, p. 93). Similar in spirit is the study by Magerman et al. (2006) where they sought to match patents to firms: they observed 74 name variations for IBM.[2]

In addition to recording errors, other common types of data collection issues include duplications, missing values, and inconsistencies. Figure 7.1 serves to highlight these errors. A duplication is highlighted in orange. Missing values are indicated in green. A suspected inconsistency is indicated in blue, because Jessica (typically a female name) is given a male attribute (Title = Mr), which is worth double-checking. Furthermore, Fig. 7.1 contains some typographical errors (such as "United Knigdom") that have not been highlighted.

---

[2] Name variations included spelling variations ("IBM" and "I.B.M."), typographical errors ("INTERATIONAL BUSINESS MACHINES"), addition of the legal form (e.g. "IBM CORP."), addition of establishment details (e.g. "IBM JAPAN") as well as other errors.

Further sources of inaccuracies in the data could come from measurement error (including biases such as self-report bias and recall bias in survey data) and statistical noise. Bias can enter datasets if the initial data sample is not representative of the wider population, as we saw in Chap. 6 on ethics. **Outliers** and extreme observations can also affect data analysis, because in skewed data a single observation can have a strong effect on the population average and the regression line of best fit. Some outliers are clearly errors and can be dropped (e.g. if a giraffe is included in a sample of puppies). Some outliers should not be dropped, even if they are very different from the other observations: an example would be lottery winners. Lottery winners may enjoy huge returns (e.g. 100 million yen) while everyone else has small losses (e.g. a 500 yen loss). However, the hope of winning is the only reason for playing. If the winner is excluded from the dataset for reasons of outlier removal, then it is hard to understand the nature of playing the lottery. A more subtle type of outlier would be a contextual outlier (Fig. 7.2). In a sample of firms, there may be a positive relationship between total sales, and number of employees. But if a firm has large sales and a tiny number of employees, this would not be an outlier when looking at the two variables individually, but would be an outlier when looking at the two variables simultaneously.

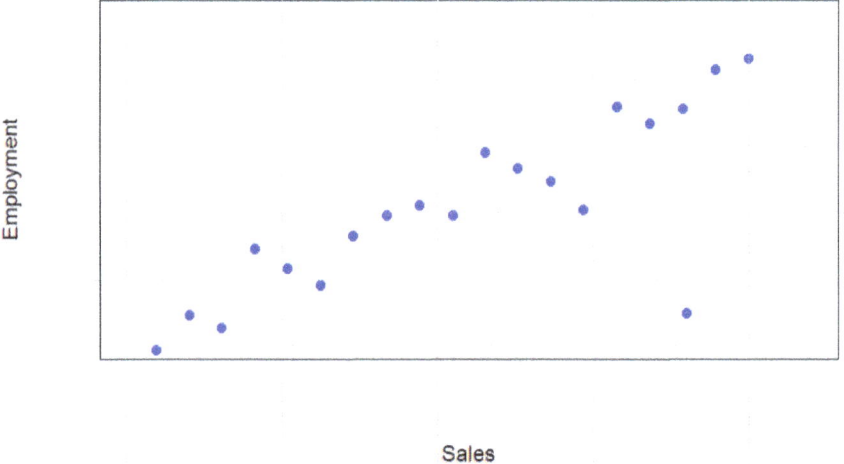

Fig. 7.2 Scatterplot of total sales, and employment, with a contextual outlier that has high sales and low employment. (*Source* Author's elaboration, see the R code file for details)

7.3 Data Pre-processing

Data pre-processing is necessary to get the data into a form that can be used for data visualization and quantitative analysis. This section discusses five common steps in data pre-processing: Aggregation, Sampling, Dimensionality reduction, Discretization, and Variable Transformation.

Aggregation refers to the act of combining objects such that they become aggregated at a level that corresponds better to theoretical or empirical interest. Aggregation can be referred to as a data reduction tool. The data become more manageable, although at the cost of sacrificing detail. One example would be the aggregation of units of time: seconds, minutes, hours, days, weeks, months, years, and so on. Another example would be aggregation in the geographical dimension: from GPS coordinate, to post code, to city, to country, etc. Aggregated data has the advantage that it offers a high-level view, with smaller statistical fluctuations, and it can be easier to process with quantitative analysis tools. However, the advantages should be considered alongside the drawbacks, which are a loss of statistical information, as well as the possible introduction of new biases.

Sampling refers to leaving out records from the data. Sampling can be a useful alternative if processing the data of the full population would be too expensive in terms of time and resources. Sampling as a pre-processing activity is different from its usual meaning in statistics, where sampling is used because it is not feasible to obtain data on the whole dataset (e.g. sampling for an election poll). An illustration of sampling as a pre-processing step is given by Hal Varian, former Chief Economist at Google: "At Google, for example, I have found that random samples on the order of 0.1 percent work fine for analysis of business data." (Varian 2014, p. 4).

Dimensionality reduction refers to the selection of a subset of the data attributes or variables. Ideally, the number of variables can be reduced without too much loss of useful statistical signal, if for example a large number of variables are highly correlated and have overlapping statistical information.

Principal Components Analysis (PCA) is a well-known data reduction tool. PCA can create a single latent variable (or small number of latent variables) to represent multiple variables that are highly correlated between them (and hence have a similar meaning because of overlapping statistical information). PCA reorganizes the data to generate a synthetic variable (or variables) that parsimoniously represents the statistical information in other variables. This could be useful in a regression setting, for example, where including many similar variables would cause problems of multicollinearity, but including one (or a small number of) PCA-generated variables leads to a simpler regression model that is potentially easier to interpret. PCA tries to prioritize the variation that is common across variables, while discarding the idiosyncratic junk. For more intuition on how PCA works, a recommended passage is James et al. (2021, pp. 253–256). Section 7.4 contains an example of PCA in R.

Discretization is a pre-processing step that transforms continuous variables into categorical/binary variables. An example was shown in Sect. 5.5, where a continuous variable (test score) was converted to a binary variable (pass or fail). There are many ways in which data can be discretized: into two or more categories; into categories on a linear or logarithmic scale; into equi-populated categories through the use of quantiles, and so on.

Variable transformation is a useful step in various situations. In cases where variables are skewed, the distribution can be asymmetric and affected by outliers, which is a problem for data visualization (e.g. if all the datapoints are squashed into a corner of a scatterplot) as well as for quantitative techniques such as OLS regression (which performs better in the case of normally-distributed error terms). A **logarithmic transformation** can help deal with these problems of skew. Taking logarithms is only possible for positive numbers; it is not possible to take a logarithm of zero, and in R this will return a missing value. One possible alternative is a $\log(1 + x)$ transformation for the variable x, such that even if x takes the value of zero, the logarithmic transformation will return a number rather than a missing value. A $\log(1 + x)$ transformation is useful for variables such as number of patents, which can be a very large integer for some firms, and zero for many other firms. The **Inverse Hyperbolic Sine** (IHS, referred to as "`asinh`" in base R) is similar to a logarithmic transformation, except that it takes not only positive but also zero and negative numbers as inputs. The IHS is useful for variables such as firm profits, which can be large and positive for successful large firms, and large and negative for unsuccessful large firms, but close to zero for many small firms. Standardization of variables is another kind of variable transformation. **Standardization** often signifies that variables are rescaled such that they all have a mean equal to zero, and a standard deviation equal to one. This procedure, which works better for normally distributed data, can make it easier to compare effect sizes across variables. Finally, reverse coding is useful in some cases where higher scores correspond to lower outcomes. An example could be a questionnaire that has two questions whose responses are negatively correlated, such as Q1: how much do you love data science, and Q2: how much do you hate applied statistics. Reverse coding the responses to one of the questions will mean that the two questions will be expected to have positively correlated responses.

Data sets in the wild often have lots of **missing values**. In some cases, with a bit of contextual knowledge, it could be relatively obvious what the missing value should be. For example, in a database that contains information on residential properties, if most houses have a missing value for private sauna, but a minority have the value "yes", then keeping in mind that most houses do not have a private sauna, we could guess that a missing value could often be replaced by "no". Taddy et al. (2023) recommend data imputation as a pre-processing technique for missing values on the explanatory variables, although data imputation should not be used for the dependent variable. If the data appears to be a sparse matrix (where the majority of cases are zeroes), it seems reasonable to replace a missing value with a zero. In other cases, a missing value could be replaced with the variable's average,

7.3 Data Pre-processing

because average values will have relatively low leverage on regression estimations. In the case of categorical variables, a missing value could be replaced with a new categorical variable called NA. For example, if we had data on the four seasons, there would now be five categories: spring, summer, autumn, winter, and NA. And finally, there is always the option just to drop the variable (or drop the missing observations) and proceed without data imputation.

Data imputation can be a controversial technique. You should be very clear about how data imputation is used, especially for scientific research and high-profile company reports. In the worst case scenario, it could look like you are faking your data to get nice results. Analysts should be transparent, to avoid accusations of engaging in researcher misconduct. One helpful piece of advice would be that, at the time when data is being imputed, a new indicator variable (here: "Imputed_ dummy"; see Fig. 7.3) can be created to indicate that the data are imputed (Taddy et al. 2023). This is useful to see if observations with missing values are somehow unusual, and allows you to check the robustness of your analysis to the data imputation approach (e.g. to see if your main results hold even when removing all cases where Imputed_dummy $= 1$).

Data preparation can raise many questions: What is the problem I want to solve? Is the data available or should I collect it myself? Are the quality and quantity of my data good enough? Which parts of the dataset are the most relevant? How do I need to pre-process my data to solve the problem efficiently? How many outliers should I remove? Data preparation is rarely something that can be done on autopilot. Instead, data cleaning can involve as many decisions as data analysis. Even in cases where research teams were given the same data and the same objective, there were large differences in data preparation procedures, definition of variables,

Fig. 7.3 Data imputation, and the creation of an "imputed" dummy. (*Source* Author's elaboration, drawing on ideas in Taddy et al. [2023, p. 124])

ID	Measurement
1	10
2	12
3	<NA>
4	8

ID	Measurement	Imputed dummy
1	10	0
2	12	0
3	10	1
4	8	0

and analysis decisions (many of which would usually not have been written up in detail), leading to different sample sizes and different results (Huntington-Klein et al. 2021). A practical piece of advice would be that data scientists should always keep a record of data preparation steps with a generous use of #comments in their R code.

7.4 R Example: PCA on Scoreboard Data

This example uses data from the European Commission's Joint Research Centre, freely available at this link: https://iri.jrc.ec.europa.eu/scoreboard/2023-eu-industrial-rd-investment-scoreboard#field_data. Then download this dataset: "**SB2023_World2500.xlsx**". (Alternatively, the data can be downloaded from this book's website).

Our task is to generate an indicator for firm size. One way of doing this could be to measure size in terms of employees, but a problem would be that some large firms (especially digital native firms) have large sales and market value, while employing a low number of employees. An alternative size indicator, besides employment, could be total sales, but a problem with this indicator is that vertically-integrated firms (that make all their inputs) might be indistinguishable from smaller firms that have large values of total sales, but relatively small values of Value Added (because much of their sales revenue is spent on buying the raw materials from outside). R&D investment is positively related with firm size for many firms, although there may be large firms with zero investment in R&D investment, such that R&D investment is not always a good indicator of firm size. More generally, we could consider that each proxy for firm size has some useful statistical information, as well as some idiosyncratic noise. Principal Components Analysis (PCA) is a useful data reduction tool, that allows us to generate a single indicator for firm size that combines information from each of the six size indicators, while discarding the idiosyncratic noise that each indicator inevitably brings.

To be precise, we generate a summary "firm size" indicator using statistical information from 6 variables: "R&D", "Net sales", "Capex", "Op.profits", "Employees", and "Market cap".[3] These 6 variables each correspond to different aspects of the underlying concept of "firm size".

We import the data, and look at some summary statistics (Fig. 7.4). A first remark is that there are lots of variables that we probably do not need (we will simplify the dataset later). Another remark is that there are lots of missing values (for example, 11 missing values, or "NA's", for the variable "Net Sales").

We therefore create a simpler dataset that is an excerpt of our 6 main variables. For PCA estimation, we use `principal` in the package {psych} (following

---

[3] These variables correspond to the following, respectively: R&D Investment, Total Sales, Capital Expenditures, Operating Profits, Employees, and Market Capitalization.

7.4 R Example: PCA on Scoreboard Data

```
> # summary statistics
> summary(sb)
      Rank             Company               Year         Country            Region           Industry         R&D (€ million)
 Min.   :   1.0   Length:2500        Min.   :2022    Length:2500       Length:2500       Length:2500       Min.   :   53.67
 1st Qu.: 625.8   Class :character   1st Qu.:2022    Class :character  Class :character  Class :character  1st Qu.:   79.27
 Median :1250.5   Mode  :character   Median :2022    Mode  :character  Mode  :character  Mode  :character  Median :  132.87
 Mean   :1250.5                      Mean   :2022                                                          Mean   :  499.77
 3rd Qu.:1875.2                      3rd Qu.:2022                                                          3rd Qu.:  297.95
 Max.   :2500.0                      Max.   :2022                                                          Max.   :37033.58
 R&D one-year growth (%) Net sales (€ million) Net sales one-year growth (%) R&D intensity (%)  Capex (€ million)
 Min.   : -69.01         Min.   :     0.0      Min.   : -100.00              Min.   :    0.0    Min.   :    0.00
 1st Qu.:   2.35         1st Qu.:   612.4      1st Qu.:    2.03              1st Qu.:    3.2    1st Qu.:   22.16
 Median :  12.46         Median :  2298.2      Median :   11.68              Median :    6.6    Median :   98.91
 Mean   :  27.07         Mean   : 10568.9      Mean   :   68.28              Mean   :  584.6    Mean   :  676.93
 3rd Qu.:  28.95         3rd Qu.:  7534.9      3rd Qu.:   23.58              3rd Qu.:   19.2    3rd Qu.:  397.53
 Max.   :13085.92        Max.   :566628.9      Max.   :61716.33              Max.   :365041.7   Max.   :39303.59
```

Fig. 7.4 Summary statistics in R. (*Source* Author's elaboration, see R code for details)

Field et al. 2012), although Taddy et al. (2023, p. 293) also recommend `prcomp` in the package `{stats}`.

The information at the bottom of Fig. 7.5 shows that first PCA component explains 62% of the total variance. In other words, one single PCA generated indicator accounts for 62% of the variation in 6 variables: not bad. The second component only accounts for 16% of the variance, and the third component even less (10%).

The score for "SS loadings" (Sum of Square loadings, also known as the eigenvalues) is 3.75 for the first component, and only 0.94 for the second component. According to Kaiser's criterion, we should only consider components with a score for their eigenvalues that is greater than 1.00, therefore we focus here only on the first component.

Nevertheless, there might be a way to improve upon 62%. Let's take a look at the data, using a useful data science technique: a scatterplot matrix (also known as a "pairs plot", Schmarzo 2016, p. 114).

```
> pc1 <- principal(sbx, nfactors = 6, rotate = "none")
> pc1
Principal Components Analysis
Call: principal(r = sbx, nfactors = 6, rotate = "none")
Standardized loadings (pattern matrix) based upon correlation matrix
                       PC1   PC2   PC3   PC4   PC5   PC6 h2        u2  com
R.D....million.       0.73  0.08  0.66 -0.09 -0.11  0.07  1  6.7e-16  2.1
Net.sales....million. 0.89  0.13 -0.28 -0.07 -0.31 -0.13  1  5.6e-16  1.6
Capex....million.     0.85  0.17 -0.12 -0.43  0.24 -0.02  1  0.0e+00  1.8
Op..profits....million. 0.82 -0.45 -0.24  0.08 -0.02  0.24  1  1.0e-15  2.0
Employees             0.62  0.68 -0.07  0.35  0.10  0.05  1 -4.4e-16  2.6
Market.cap....million. 0.80 -0.47  0.13  0.26  0.14 -0.18  1  0.0e+00  2.1

                       PC1   PC2   PC3   PC4   PC5   PC6
SS loadings           3.75  0.94  0.61  0.39  0.20  0.12
Proportion Var        0.62  0.16  0.10  0.07  0.03  0.02
Cumulative Var        0.62  0.78  0.88  0.95  0.98  1.00
Proportion Explained  0.62  0.16  0.10  0.07  0.03  0.02
Cumulative Proportion 0.62  0.78  0.88  0.95  0.98  1.00
```

Fig. 7.5 inspecting the 6 PCA components in R. (*Source* Author's elaboration, see R code for details)

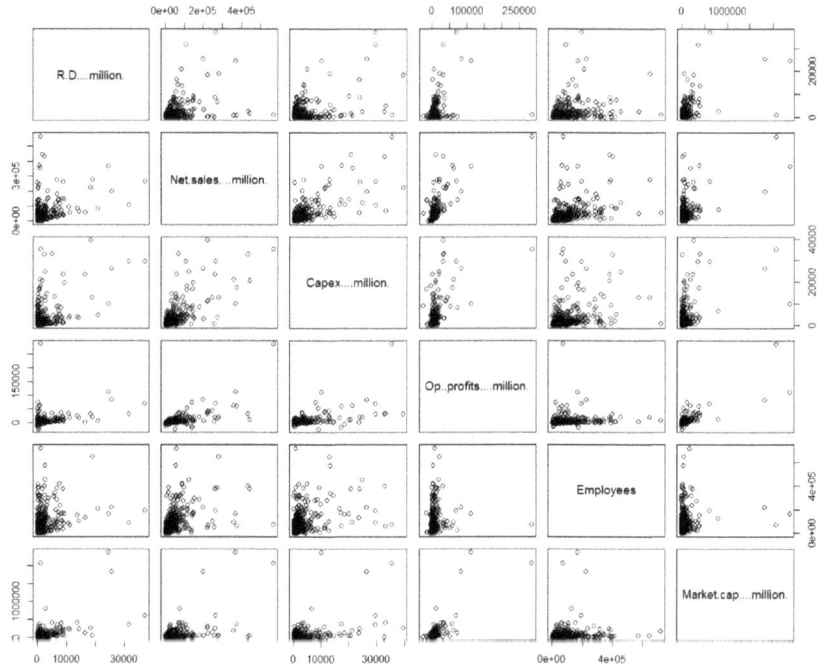

Fig. 7.6 scatterplot matrix in R. (*Source* Author's elaboration, see R code for details)

The scatterplot matrix in Fig. 7.6 shows the pairwise relations between these 6 variables. Here we see that most of the datapoints are bunched up in the bottom left corner of their respective graphs, indicating that the data are highly skewed. As such, transforming the data, using a logarithmic transformation, will help to smooth out the datapoints across the relevant range. The next step in the R code is to repeat the scatterplot matrix after taking logarithms of the variables, but this gives some red ink (an error message) alongside the new graph.

The problem is that it is only possible to take logarithms of positive numbers. If a variable (such as Operating Profits) contains zeroes or negative values in some cases, it will not be possible to take logarithms, and missing values (NA's) will be created. Figure 7.7 shows that taking logarithms means that we move from 2487 to 1745 observations for Operating Profits.

Therefore, we try an alternative model. Model 2 drops the problematic variables,[4] which sidesteps the problem, although a drawback is that we have only 3 variables (instead of 6) going into our PCA. The first PCA component explains

---

[4] Missing values affected the variables Operating Profits, and also Net Sales and Capital Expenditures.

7.4 R Example: PCA on Scoreboard Data

```
> colSums(!is.na(sbx))
        R.D....million.    Net.sales....million.    Capex....million.  Op..profits....million.              Employees
                   2500                     2489                 2437                     2487                   2344
    Market.cap....million.
                   2320
> colSums(!is.na(log(sbx)))
        R.D....million.    Net.sales....million.    Capex....million.  Op..profits....million.              Employees
                   2500                     2489                 2437                     1745                   2344
    Market.cap....million.
                   2320
```

Fig. 7.7 Missing values in the data. (*Source* Author's elaboration, see R code for details)

74% of the cumulative variance (compared to 62% for Model 1), but this is less impressive because there are only three variables included in the PCA.

Another technique for transforming the data is the Inverse Hyperbolic Sine (IHS), which resembles the logarithm for positive values, but can also reduce the skew over the negative range, and can also generate a number for the value zero. The IHS is useful for variables such as Operating Profits, which can be positive and large, or equal to zero, or even negative and large (e.g. for an unprofitable large firm). Model 3 attempts a PCA with an IHS transformation, and the first PCA component explains 69% of the variation. Can we improve upon this? Let's take another look at the data, using another scatterplot matrix (Fig. 7.8). Here we see a bizarre phenomenon affecting the Operating Profits variable: there is something like a "V"-shaped cloud of points indicating a non-linear relationship with other variables (in the fourth column of the scatterplot matrix). This seems bizarre at first, but there is an explanation. Firms with large positive profits are probably large. Firms with profits close to zero are probably small. But then, firms with large negative profits are probably large (small firms would not be allowed to make billions in losses.) Hence, the nonlinear relation occurs because large firms make either huge profits or huge losses, but are rarely found in between these two extremes (with small firms occupying this middle part). As such, it would be better to take logarithms of the absolute size of profits. Furthermore, a useful trick is to add a "grain of sand" into the logarithmic transformation, i.e. to take $\log(1 + x)$, instead of $\log(x)$, so that the logarithm can be calculated even if $x = 0$. Model 4 shows that the first PCA component explains 72% of the common variance in the 6 variables included in the model, which is satisfactory for our purposes.

This example focused on applying PCA to data on firm size and performance. PCA is also often applied for questionnaire data, where there are many similar questions, each shedding light on slightly different aspects of the same underlying factors. With questionnaire data, there are often several latent dimensions (not just one), suggesting that we would be interested in more than 1 of the PCA-generated components.

PCA is also useful for social science research which focuses on latent variables that cannot easily be measured, such as intelligence, personality, or leadership ability. A number of survey questions could shed light on different aspects of the same underlying phenomenon, with PCA being used later on to synthesize a summary indicator based on these survey questions.

PCA is an example of what data scientists would call unsupervised learning: PCA will return some results, but we are not sure what the results mean or how to

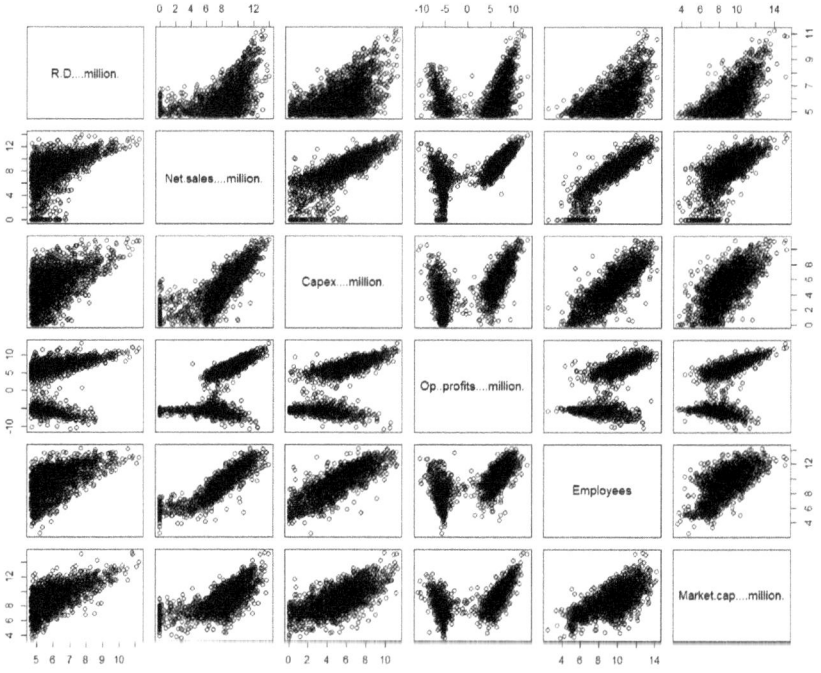

Fig. 7.8 Scatterplot matrix in R. (*Source* Author's elaboration, see R code for details)

interpret them, and we are exploring and looking for patterns rather than predicting an outcome.

Further Reading

The books by James et al. (2021) and Taddy et al. (2023) have chapters that focus on PCA in R software. Field et al. (2012) contains examples in R of applying PCA to questionnaire data, to uncover the underlying latent variables that represent the common variation in the survey responses.

References

Davenport, T. (2014). Big data at work: dispelling the myths, uncovering the opportunities. Harvard Business Review Press: Cambridge, MA.
Field, A. Miles, J., & Field, Z. (2012). Discovering statistics using R. SAGE publications Inc, Thousand Oaks, California, 91320, USA. https://sloanreview.mit.edu/article/seizing-opportunity-in-data-quality. November 29.

References

Huntington-Klein, N. (2021). The Effect: An Introduction to Research Design and Causality. Chapman and Hall/CRC: New York, USA. https://doi.org/10.1201/9781003226055. Free to read online here: https://theeffectbook.net.

Huntington-Klein, N., Arenas, A., Beam, E., Bertoni, M., Bloem, J., Burli, P. H., Chen, N., Grieco, P. L., Ekpe, G., Pugatch, T., Saavedra, M. H. (2021). The influence of hidden researcher decisions in applied microeconomics. Economic Inquiry, 59(3), 944–960. https://doi.org/10.1111/ecin.12992

Kenett, R. S., & Redman, T. C. (2019). The Real Work of Data Science: Turning data into information, better decisions, and stronger organizations. John Wiley & Sons.

Redman, T. (2017). Seizing opportunity in data quality. MIT Sloan Management Review.

Schmarzo, B. (2016). Big Data MBA: Driving business strategies with data science. John Wiley & Sons.

Segalla, M., & Rouziès, D. (2023). The Ethics of Managing People's Data. Harvard Business Review, 101(4), 86–94.

Taddy M., Hendrix L., Harding M. C. (2023). Modern Business Analytics. Practical Data Science for Decision Making. McGraw Hill, New York, NY.

Varian H.R. (2014). Big Data: New Tricks for Econometrics. Journal of Economic Perspectives 28(2), 3–28.

The User Experience (UX) 8

Focusing on customer needs and improving the user experience is a classic theme from the business world, with recommendations coming from the literature on digital transformation (Rogers 2016) as well as the literature on the lean startup approach (Ries 2011). Successful innovators need to stay focused on customer needs and the user experience. This is especially true in the digital age. Big data and analytics enhance the promise of UX, because now there are all kinds of useful data-driven insights. Big data also allows firms to analyze a variety of UX-related metrics, such as customer engagement, conversion rates of customer to higher value categories, and retention rates (or turnover) of customers. Emerging technologies such as cloud services, Internet of Things (IoT) sensors, new devices, and wearables are also giving new opportunities for improving UX and learning about how to better meet customer needs.

The digital age also brings opportunities for firms to mess things up. There is so much data and information around, it is easy to flood consumers. In the age of abundant data, the imperative is to present customers with the smallest amount of highly-relevant information that they need to make a decision. This is discussed in more detail later.

8.1 Customers are Spoilt

Customers these days are spoilt (Vaz 2021). Customers who have a seamless experience in one area are often quickly frustrated when things are not quite so smooth in other domains. For example, one digital company may offer a seamless experience such that customer data is only entered once, or maybe not entered at all if the data is passively collected. Then, when a customer goes to a different sector/industry (such as a bureaucratic administrative task), they might have long waiting times, in-person procedures rather than online, being repeatedly asked the same questions, and filling in the same data multiple times for multiple departments. Customers do not enjoy filling in forms with their data many times. Also, customers do not like making lots of decisions when firms can already draw on their customer records to make suggestions based on past customer preferences. To succeed in the modern era, firms need integrated systems. Firms need to have a common data repository (i.e. a data lake) that spreads across the organization rather than being divided into siloes.

A consumer's experience in one context affects the consumer's expectations elsewhere. Consider the case of an individual on a business trip. The journey starts with an Uber driver, who knows where to find you, and where you want to go, and the price is clearly indicated. Next, arriving at the airport, the airline may already know your preferences regarding where you would like to sit, and what meal you would like to eat, and can offer you a fast online check-in and fast-track movement through the airport. Then, the same individual arrives at a traditional hotel, where they have to wait at a crowded reception, get checked into multiple different systems for different activities, and then get allocated a room based on the hotel's availability (rather than the customer's preferences). This leads to frustration on the part of the consumer, who has otherwise had a seamless experience that day.

Organizations used to think about data in terms of departments and functions, such as customer relationship management (CRM), online sales, marketing, finance, and so on. Siloed data and multidivisional corporate structures worked well in the previous business age, but are problematic in the digital age. Data needs to be connected across functions, and this requires the appropriate data infrastructure. Firms need to be able to "connect the dots" across the organization to serve customers' needs (Vaz 2021).

Despite some success stories about improvements in the user experience, bureaucracy can sometimes be despairing, and maybe one day we may joke about our previous frustrations. There are several humourous examples making light of awful user experiences, such as products with uncomfortable designs[1] and awkwardly-designed online interfaces.[2]

8.2 The Value of Customer Insights

Building a system of privileged insights with your customers can be extremely valuable. The case of Adobe is interesting.[3] For a long time, Adobe used to sell its software (such as Photoshop, Illustrator, and InDesign) in the form of CDs through third-party sellers. Adobe was in a growing industry, and its performance was fairly good, because its software products were popular. However, Adobe's

---

[1] See https://www.theuncomfortable.com/ [last accessed 8th July 2025].
[2] See https://userinyerface.com/ [last accessed 8th July 2025].
[3] See Cox (2019) and Leinwand and Mani (2022); see also https://business.adobe.com/uk/customer-success-stories/adobe-experience-cloud.html [last accessed 16th July 2025].

leaders wondered whether its growth could be accelerated. A major problem was that Adobe had little contact with customers. Its products were mainly sold in the form of CD packages and used offline, and the only contact that Adobe had with customers was at the time when they registered their newly-purchased product. Adobe therefore knew little about its customers and how they actually used Adobe products. Then came the DDOM: Adobe's Data Driven Operating Model. Adobe shifted away from CDs to a cloud-based Software-as-a-Service (SaaS) arrangement, based on a direct subscription revenue model. Adobe also embarked on a company-wide initiative to accelerate its digital transformation and to develop data architectures that would allow it to better observe and leverage customer insights. A change in culture occurred, from decision-making based on intuition and educated guesses, to data-driven decision-making. Adobe could now observe the full customer journey, not just at the product registration stage, but at the stages of discovery, trial, purchase, usage, and renewal. New KPIs were set up to measure and enhance the full customer experience in both financial and non-financial terms. As a result, Adobe could see how customers were actually using their products, in real time. Adobe could observe the pain points and problems faced by customers, that in some cases were preventing customers from realizing the full potential of Adobe software. It is likely that some customers wanted to stop using Adobe products, not because they were inadequate, but because the users could not figure out how to use the products properly, and were not aware of the products' true possibilities. Until the shift to a cloud-based SaaS business model, Adobe was unaware that many customers were using their products incorrectly. Now, Adobe could observe which activities were causing problems for users, and work on improving the experience for these activities. Adobe could even offer in-app real-time suggestions, to provide assistance to users who were facing difficulties. For example, if a customer was becoming frustrated while editing a photo, Adobe could infer their problems (using information on which menus they accessed and where they clicked), and send messages such as "Hey, I think you're trying to apply this filter." Adobe also observed that some neglected products were actually driving enormous value for customers, although Adobe had previously prioritized other products because they instinctively felt right. This led to impressive results in terms of financial performance: "Adobe's leaders credit most of the company's revenue growth from $5.9 billion in 2016 to $12.9 billion in 2020 to their data-driven insights capability" (Leinwand and Mani 2022, p. 96).

8.3 UX and Innovation

Better understanding customer needs also has advantages in terms of enhancing the efficiency of innovation activities, in terms of shortening the innovation cycle, and in terms of producing what customers truly want. Idea generation can occur faster, if firms can draw on the insights and suggestions of customers, as well as observing how customers use their products. Prototypes can be tested at a relatively early stage (with basic prototypes that have a minimal functionality), to learn from rapid experimentation and to be early to shut down unattractive avenues.

Regrettably, there have been too many cases of products that are launched on the market, only to find out that they are not what customers need. For example, the firm may have designed a square thing, while customers would have preferred a round thing. Firms may have packed their new product with a large set of features that most customers may never use. Better understanding of UX could have prevented the situation where firms only learn about what customers actually want at the time of the market introduction of an unappealing product.

Figure 8.1 highlights how rapid experimentation can lead to shorter innovation cycles. Digital firms seek to "learn early" by validating a prototype (the "Minimum Viable Product" or MVP) with customer feedback. If the customer response suggests that the project is misguided, or if certain assumptions about customer pain points are not supported, the project can pivot into a different strategic direction. We are painfully aware that too many innovations get to market, only to find that customers do not need this new thing. "Fall in love with the problem, not the solution" is a useful proverb because, all too often, innovators can be awestruck by a certain shiny new technology (such as blockchain) and try to repackage it somehow into a new product offering, and lose track of the importance of focusing on what customers really need. Failure is not considered to be problematic if it occurs early (before too much time, money, and other resources have been spent on the project), especially if it leads to valuable learning and allows the team to pivot towards a more promising direction. Failure does not necessarily have catastrophic connotations, as explained by Thomas Edison: "the real measure of success is the number of experiments that can be crowded into 24 h." Another quote from Edison is that "I have not failed. I've just found 10'000 ways that won't work." (Kane et al. 2019, p. 206). Indeed, failed experiments can be a useful stock of knowledge in form of trade secrets, which are an economically important category of intellectual property.

Our discussion of innovation highlights how digital transformation requires complementary changes in the corporate culture. DX leads to the use of low-stakes experimentation (A/B testing) as a part of routine operations. In such an environment, decision-making moves from a traditional style which might involve bowing before seniority and following their gut-feelings, towards a culture of testing and validating, and respect for data. Testing ideas becomes cheaper and faster, experiments are conducted continuously, and everyone can get involved. Employees at various levels of the organization are encouraged to explore their hunches, and the data decides who wins.

Digital Transformation also encourages a deeper collaboration between R&D and IT (Porter and Heppelmann 2015; see their section "Implications for Organizational Structure"). Traditionally, R&D was about products, while IT provided the infrastructure to support bureaucratic processes such as Enterprise Resource Planning (ERP) and Customer Relationship Management (CRM). Recently, however, IT hardware and software are embedded in a vast range of products. Also, IT plays a crucial role in generating ideas, designing prototypes, running simulations

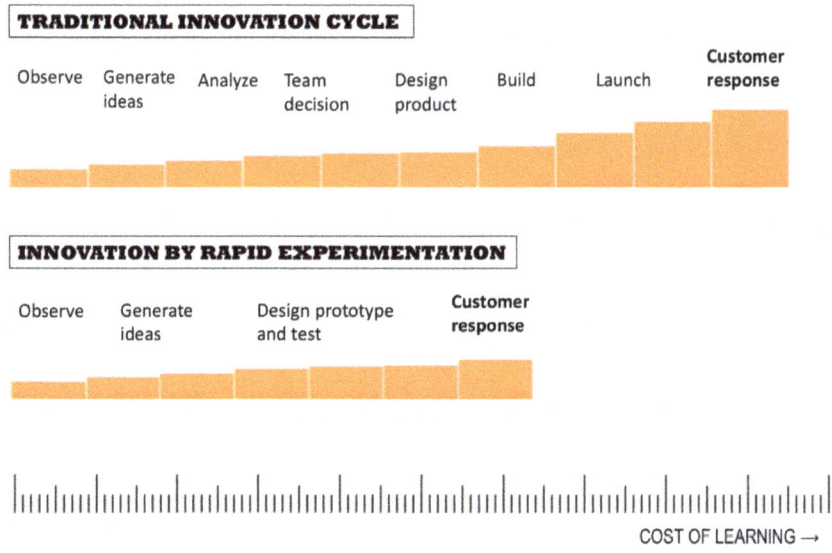

Fig. 8.1 DX shortens the innovation cycle. (*Source* Author's elaboration, inspired by Rogers [2016])

and experiments, gathering feedback from users, and so on. As such, DevOps (Software Development and IT Operations) are an essential part of successful innovation teams, helping to shorten product-development cycles.

8.4 Presenting Relevant Actionable Data with Dashboards

In the digital age, there is the promise to make better decision-making based on available data. The challenge is to get the right data to the right people, at the time when they make decisions. The data need to be cleaned, processed, and delivered in a targeted way. If decisionmakers want to drill down, and study the details, this option should also be available to them. It is important to avoid flooding decisionmakers with too much information, and irrelevant information.

Schmarzo (2016) gives an example of a user experience in the domain of mobile phone subscriptions. An individual receives a notification by email that they are near the end of their monthly data plan limit. However, this email does not give information on how much of the data plan is left in the current month, or when the next billing period starts. What customers what to know is: at the current rate of data usage, when will you be likely to exceed the monthly limit. If this information is not given, it is not useful for the user. In the worst-case scenario, firms may be sending threatening emails (e.g. you may be liable for penalty fees and surcharges) in cases where such threats are not relevant (e.g. it is very unlikely that you will exceed your data plan limit, if you have only used 70% of your monthly data plan, and your new monthly billing period starts in 2 days). Much more useful would be to let the customers know how much time has elapsed in the current month, how much data has been used, and the predicted total use by the end of the month. Other useful information would be advice on which mobile phone apps to shut down (e.g. GPS tracking for map apps) in order to save on data usage. It would also be useful to give recommendations for possible actions to take, such as adding a temporary booster to the data plan, or making a permanent upgrade to the monthly data package if data usage patterns in previous months suggest that this could be useful.

Figure 8.2 gives an example of a consumer-facing dashboard, in the context of a laundry smartphone app. Traditionally, the user experience was inconvenient: for lack of knowledge, users would have to visit the laundry facility in person, only to find that their favourite machines are all being used, or (even worse) find that the wash cycles are finished but previous customers have not removed their washed clothes. The dashboard gives useful information regarding the total number of machines and their capacity, the number of available machines, whether they are functioning (or broken), and also the time remaining for machines that are being used. This knowledge is actionable in terms of telling consumers whether to drag their clothes to the laundry facility or wait at home; and how much time is remaining for their wash programme.

Dashboards can therefore be a useful way of presenting the data needed to make informed decisions. Dashboards can enrich the user experience by presenting the right information (or actionable insights) to the right user to make the right decisions at the right time. Dashboards should be designed with a keen awareness of customer needs, and the information customers need to make good decisions. This will probably involve a mix of the presentation of historical data (backward-looking data reports) as well as predictions about the future (expected changes that come from predictive models, as well as an indicator of their uncertainty), and prescriptive recommendations (actions to take on the basis of expected changes). Dashboards should also be designed in such a way that the main messages can easily be grasped by the reader. This means that the basic principles of data visualization (discussed in the next chapter) should be kept in mind when setting up the dashboard.

Dashboards can be prepared to improve decision making by all stakeholders. Dashboards are well-known in the case of customer dashboards, but dashboards can also be designed for other stakeholders, such as empowering frontline employees (for example providing recommendations for inventory and promotions for store managers in a retail chain), as well as for channel partners (providing information to suppliers regarding requirements and logistics).

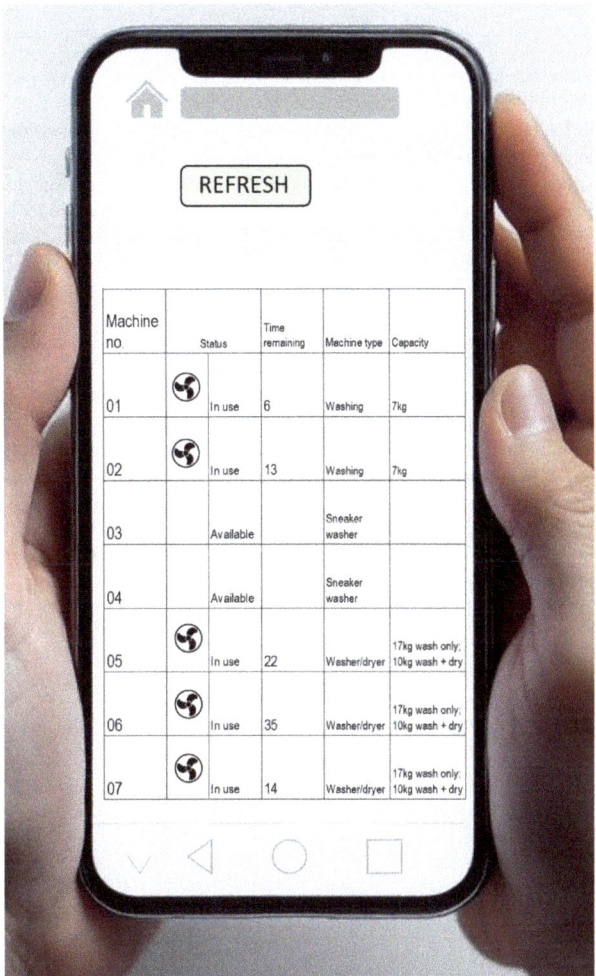

Fig. 8.2 Example of a user-facing dashboard, for a laundry smartphone app. (*Source* Author's elaboration, inspired by https://laundry.senkaq.com/, with graphical help from mistral.ai)

Further Reading

Wexler et al. (2017) and Wells and Chiang (2017) focus on how dashboards can improve the user experience by providing actionable knowledge.

References

Cox, E. (2019). How Adobe Drives Its Own Transformation. 16th May, 2019. https://blog.adobe.com/en/publish/2019/05/16/how-adobe-drives-its-own-transformation

Kane, G. C., Phillips, N., Copulsky, J. R., & Andrus, G. R. (2019). The Technology Fallacy: How people are the real thing to digital transformation. MIT Press.

Leinwand, P., & Mani, M. M. (2022). Beyond Digital: How Great Leaders Transform Their Organizations and Shape the Future. Harvard Business Review Press: Cambridge, MA.

Porter, M. E., & Heppelmann, J. E. (2015). How smart, connected products are transforming companies. Harvard Business Review, 93(10), 96–114.

Ries, E. (2011). The lean startup: How today's entrepreneurs use continuous innovation to create radically successful businesses. Crown Books: New York, USA.

Rogers, D. L. (2016). The digital transformation playbook: Rethink your business for the digital age. Columbia University Press.

Schmarzo, B. (2016). Big Data MBA: Driving business strategies with data science. John Wiley & Sons.

Vaz, N. (2021). Digital business transformation: How established companies sustain competitive advantage from now to next. John Wiley & Sons.

Wells, A. R., & Chiang, K. W. (2017). Monetizing Your Data: A Guide to Turning Data into Profit-Driving Strategies and Solutions. John Wiley & Sons.

Wexler, S., Shaffer, J., & Cotgreave, A. (2017). The big book of dashboards: visualizing your data using real-world business scenarios. John Wiley & Sons.

Data Visualization

9.1 The Importance of Graphs

Graphs are an important way of communicating the results of data analytics, and in many cases graphs can be the most efficient way of helping us to understand data. For example, it is much easier to see patterns in the data (trends, periods of unusual volatility, seasonal patterns, existence of outliers) if you have 100 datapoints in a time series plot, compared to having the 100 observations presented in a table of numbers. In the previous chapter, we saw that graphs are important for UEX dashboards, to quickly convey information in an efficient way, and to quickly provide an overview of the current situation to decision makers such as high-level executives.

9.2 Data Visualization Principles

Bad graphs are all around us. Bad graphs have misleading axes, confusing crosshatching effects, annoying colour schemes, and often confuse the reader as the eye jumps from graph to legend and back to graph. Some bad graphs are noteworthy for their comedy value.[1] In many cases, bad graphs are designed to deliberately trick the reader into misunderstanding things, for example if there is a mismatch between a number's size and the quantity of ink used to represent it (Tufte 2001).

Supplementary Information The online version contains supplementary material available at https://doi.org/10.1007/978-981-95-2433-4_9.

[1] See for example https://viz.wtf/ and https://www.buzzfeednews.com/article/katienotopoulos/graphs-that-lied-to-us [last accessed 25th July 2025].

Good graphs, however, can be highly effective. In terms of the efficiency of data representation, good graphs can present more than 10,000 numbers per square centimeter on a printed page (Tufte 2001, p. 166) which is vastly superior to what can be achieved with a table of numbers. Good graphs can be designed with careful consideration of a small number of basic principles of data visualization, which are discussed in the following subsections.

9.2.1 Basic Principles of Data Visualization

Four basic principles of data visualization can help convert a mediocre graph into an excellent graph: simplicity, clarity, avoiding repetition, and emphasizing your message.

Simplicity refers to letting the data speak in a clear way, without encumbering the graphic with unnecessary ink. Tufte (2001) measures the simplicity of a graph using the data-ink ratio, defined in this way:

$$Data - ink\ ratio = \frac{data - ink}{total\ ink\ used\ to\ print\ the\ graphic}$$

The data-ink ratio measures the proportion of a graphic's ink that is devoted to the non-redundant display of useful quantitative information. Graphs should use ink for the purpose of representing the data, and non-data-ink such as axes, legends, gridlines, unnecessary repetitions, and 3D effects should be minimized. The enemy of simplicity is **"chartjunk"** (Tufte 2001, Chapter 5) which is ink used to decorate a graph, but without providing the reader with any new statistical information. Chartjunk is associated with a low efficiency in the use of ink to represent quantitative information. As with a first draft of a written text, graphs can often be improved by further rounds of editing down to succinctly present the data in a simple and powerful way.

Clarity refers to presenting the data in such a way that the viewer can quickly grasp what the variables are and get a feel for the raw patterns in the data.

Avoiding repetition relates to how graphs should strive to be as efficient as possible.

Linking back to the data-ink ratio, redundant ink should be ruthlessly trimmed down and edited out.

Finally, an important principle is to **emphasize your message**, rather than leaving readers wondering what they are supposed to be looking at. For example, if you have a scatterplot with 100 datapoints, corresponding to 100 countries, you do not need to label all of the datapoints if you want to draw the viewer's attention specifically to a handful of countries such as Japan, China and South Korea. Annotations can be used on graphs to give helpful context, to highlight datapoints of key interest, or to indicate key events such as regime changes in a time series graph. Annotations on graphs are not bad practice. A suitable annotation is useful

9.2 Data Visualization Principles

information that helps the reader to know where to look, and can count as data-ink (as opposed to non-data-ink) and thereby help to improve the data-in ratio shown above (Tufte 2001).

Weaving these four principles into the visual representation of quantitative data is not just a matter of computation, but calls for creative thinking:

> It takes imagination to create high-quality images that illustrate data accurately and effectively and also show some understanding and appreciation of how people acquire information. (Schwabish 2014, p. 210)

9.2.2 Pre-attentive Processing

Well-designed graphs can tap into the instinctive shortcuts used by the brain to process information, in order to quickly draw the reader's mind to the most important datapoints or trends in the graphic. Pre-attentive processing therefore plays an important role in the efficient communication of quantitative information in graphs. Pre-attentive information refers to the features in a graph that are processed by our brains before we understand the graph at a conscious level. Colour is perhaps the most important technique for grabbing the viewer's attention to a specific part of a graph, although other pre-attentive features are also worth mentioning because they can also be effective. Figure 9.1 presents a number of the most common pre-attentive features: colour hue, colour intensity, size, shape, added marking, and spatial position.

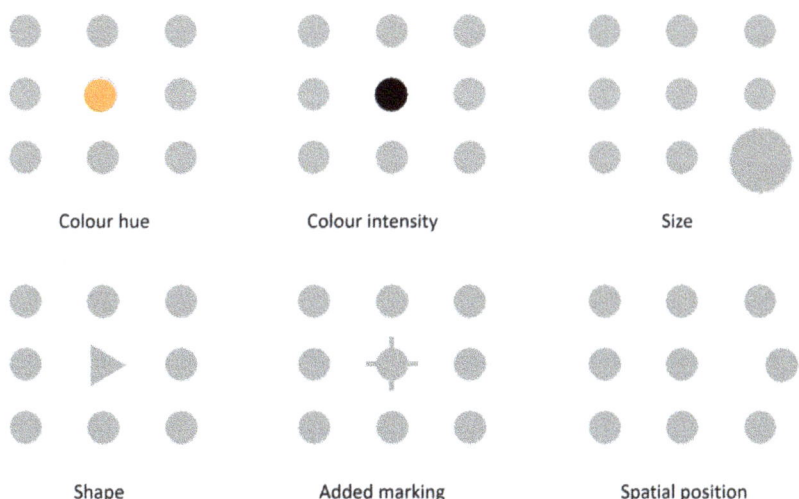

Fig. 9.1 Common features for pre-attentive processing (*Notes* Author's elaboration, similar in style to Nussbaumer Knaflic [2015, Fig. 4.4] and Wexler et al. [2017, Fig. 1.10])

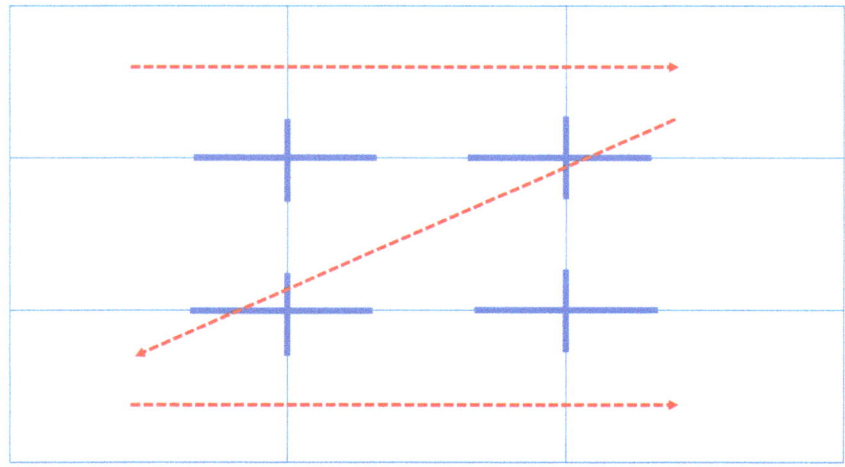

Fig. 9.2 The rule of thirds, and the Z-pattern of scanning (*red dashed arrows*) (*Source* Author's elaboration)

9.2.3 Further Principles of Visual Design

Two other principles of visual design are worth mentioning: the rule of thirds, and the Z-pattern of scanning (Wells and Chiang 2017). These are represented in Fig. 9.2. **The rule of thirds** is a visual ordering concept that is widely used in the film and photography domain. The visual area is divided into a 3×3 matrix, and the gridlines from this matrix are used to assign positions to visual objects according to their relevance.

The **Z-pattern of scanning** can build on the rule of thirds layout to give useful insights into how information is gathered. The Z-pattern of scanning is an intuitive insight into information display that is based on how individuals read pages.[2] The Z-pattern of scanning suggests that we first scan information on the top row, from left to right, until we get to the end of the page. Then, the reader's eye scans down from the top right corner through to the central part of the area down to the bottom left-hand corner, until it proceeds to scan along the bottom row until it gets to the bottom right-hand corner. The Z-pattern of scanning would suggest that, for newspapers for example, the most important stories should be at the top-left and top-right corners, with stories at the left and right side of the central row being more susceptible to being overlooked. Important supporting information can be placed in the middle of the diagonal, and the bottom line can contain conclusions, recommendations, and calls for action.

---

[2] The Z-pattern of scanning is often referred to in English language publications, where reading proceeds from top-left to top-right. The Z-pattern of scanning may be less relevant in other cultures where reading proceeds in a different way, such as from right to left, or from top to bottom.

Both of these principles (the rule of thirds, and the Z-pattern of scanning) are relevant for designing dashboards to present graphical information to users in an efficient and intuitive way.

In the context of data dashboards, the Z-pattern of scanning has implications for where to put information such as switches allowing users to toggle over different data categories (e.g. to filter for years, or switch between geographical areas). In cases where users are not familiar with a dashboard, it would be better to position such switches and filters at the top left corner, where this important information will be noticed. In cases where users are already familiar with using the dashboard, however, such switches and filters can be positioned at the right-hand side of the middle row, without having to worry about whether these useful features will go un-noticed by users (Wexler et al., 2017).

9.3 Some Common Types of Graph

9.3.1 Pie Charts

Pie charts are frequently shown in scientific publications and business reports—and probably far too frequently. Pies are not good for you. Pie charts are widely considered by data visualization experts to be an unhelpful way of presenting data. Healy (2018, p. 26) states that "pie charts are usually a bad idea." Schwabish (2014, p. 222) explains that "because pie charts force readers to make comparisons using the areas of the slices or the angles formed by the slices—something that our visual perception does not accurately report—they are not an effective way to communicate information." For those who are sick of seeing charts and numbers, a pie chart could offer some relief. However, if the reader is sick of seeing charts and numbers, they should maybe just close the report and do something else instead. Graphs are intended for showcasing interesting data. It is not recommended to embellish a graph with fancy effects to make it more appealing: the data itself should be appealing. If the numbers are not interesting, this probably means that you are showing the wrong numbers. If the numbers are interesting, then the awkward way that pie charts present visual information will be a source of annoyance rather than a valuable method for data visualization.

Figure 9.3 gives an example. On the left, we have a pie chart (even worse, a 3D pie chart, where the 3D effect distorts the relation between amount of ink used for each category and the overall importance of each category). Some readers may be distracted by the "flashy" use of computer-generated graphics: in particular, this would have been an impressive display of computer power if we saw this graph 30 years ago, although it is less impressive these days. However, it is difficult to compare the relative sizes of the categories. Is B larger than E? What about C and D? A simple bar chart (shown in Fig. 9.3 right) would be much more informative, because the human eye performs better at comparing lengths compared to comparing pie slices. With Fig. 9.3 right, answering the same questions (Is B larger than E? What about C and D?) is immediately clear.

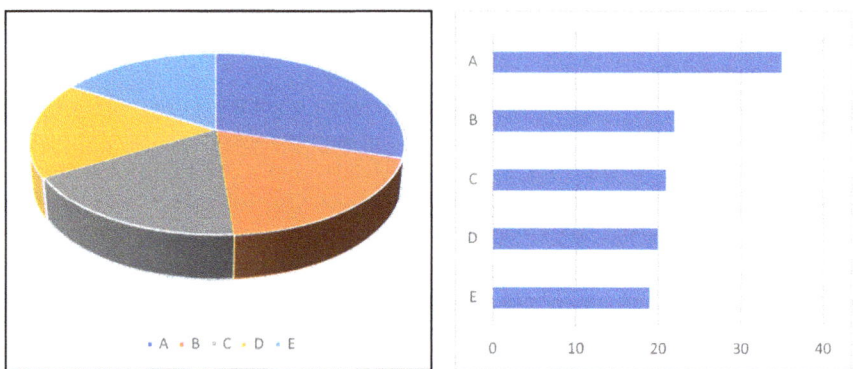

Fig. 9.3 Graphs in Microsoft Excel. A standard 3D pie chart on the left, and a horizontal bar chart showing the same information on the right (*Source* Author's elaboration)

9.3.2 3D Plots

A 3D plot is another example of poor data visualization (Healy 2018; see also the example in Fig. 9.4). The graph contains too much non-data-ink, such as gridlines and 3D effects. The 3D effect even interferes with our sense of spatial perspective, such that it harder to read off the values of the heights of the bars than if the bars were given a 2-dimensional representation. A 2D plot would be better than a 3D plot, but a simple table could even be preferable to a 2D plot. "Tables are preferable to graphics for many small datasets." (Tufte 2001, p. 178). With so few datapoints, the graph fails to leverage the strengths of data visualization.

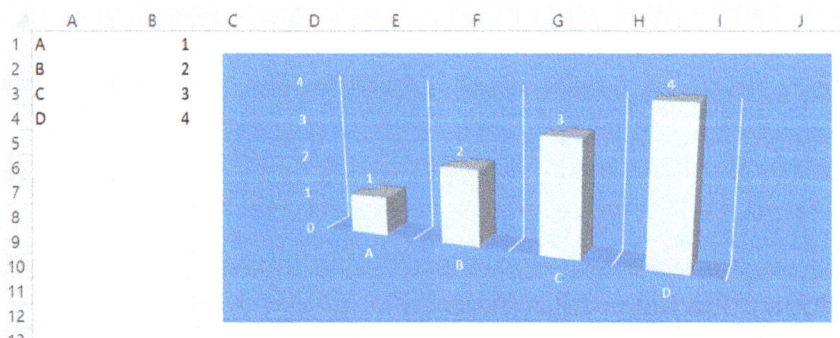

Fig. 9.4 3-Dimensional plot in Microsoft Excel (*Source* Author's elaboration)

9.3.3 Overplotting and Some Remedies

This subsection involves a fictional example in R, set in the context of the scores for a product review given by 2 online raters, who each have to give an integer score ranging from one star to five stars. The R code file first generates this dataset, then shows the scores in a scatterplot (Fig. 9.5), before further exploring the data.

To begin with, Fig. 9.5 shows the data using a scatterplot. Do the scores look related? It is not entirely clear, but probably we could expect a positive relationship between the scores of rater1 and rater2, particularly if we forget about the case of {rater1 = 5, rater2 = 1} which could plausibly be a simple outlier. If the relationship is positive, we would expect a positive correlation coefficient. However, the correlation coefficient is calculated to be strongly negative, at − 0.634. Furthermore, the regression coefficient indicates that a unit increase in the score of rater1 is associated with a decrease of −0.775 in the score of rater2. The negative relationship becomes even clearer when we plot the regression's fitted line on top of the graph. How can the relationship be so strongly negative, when a first look at the scatterplot suggested a positive relationship?

The answer is that there are some datapoints plotted on top of each other. This is especially common in cases where the variables are not continuous, but the data are forced to take a small number of outcomes, such as with binary variables, or integer variables like we have here in the ratings data. There are several ways to become aware of the problem of overfitting. One way, shown in Fig. 9.6 (left), is to use the **jitter** option. When we jitter the datapoints, a small amount of random

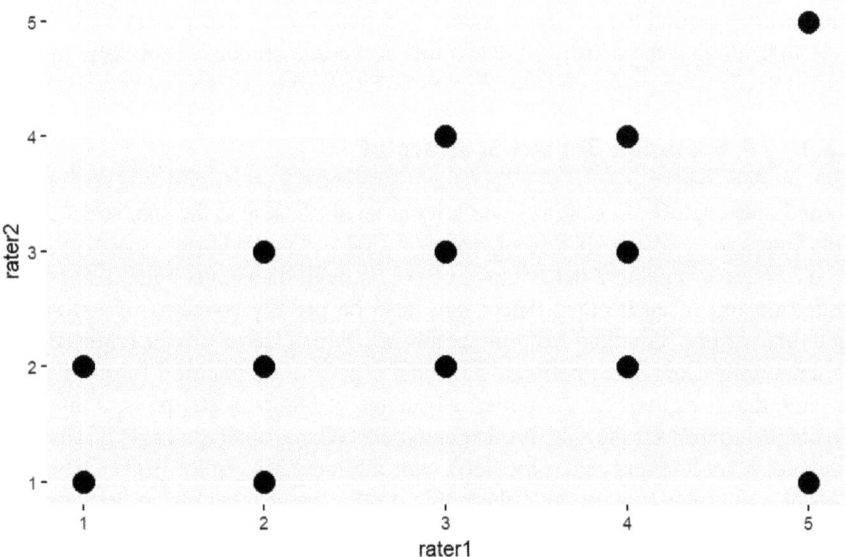

Fig. 9.5 Scatterplot in R (*Notes* Author's elaboration, see R code file for details)

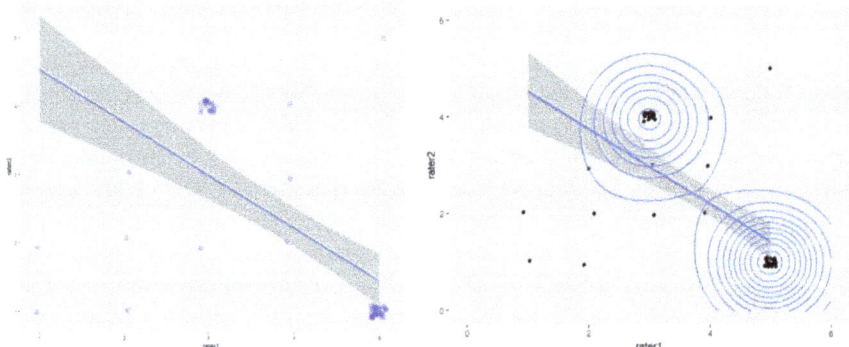

Fig. 9.6 Both the jitter option with semi-transparent datapoints (*left*) and the contour plot (*right*) reveal the phenomenon of overplotting (*Notes* Author's elaboration, see R code file for details)

noise is added to the x coordinates and y coordinates, such that the datapoints are spread out slightly rather than being plotted exactly on top of each other. In addition to the jitter effect, we have set "alpha = 0.2" to make the datapoints not be fully opaque, but to have 20% transparency. This is an additional way of showing when datapoints are being plotted on top of each other. Another way to show overfitting could be to use a contour plot (sometimes referred to as a heat map, see Fig. 9.6, right). A **contour plot** uses contour lines to indicate the intensity of datapoints on the two-dimensional graphical surface, in the same way that contours on a geographical map indicate the altitude of hilly terrain. The contour lines around the points of {rater1 = 5, rater2 = 1} and {rater1 = 3, rater2 = 4} indicate that the density of datapoints at these positions is especially high.

9.3.4   R Example: Binned Scatterplot

Binned scatterplots are gaining popularity as an alternative to the standard scatterplot (Starr and Goldfarb 2020; Cattaneo et al. 2024). The traditional scatterplot has the drawback of having a dense cloud of points, with datapoints being sometimes plotted on top of each other. There may also be privacy concerns, if individual datapoints can be identified from the graph, which has led to stricter requirements for anonymity from some national statistical offices as a condition for publishing research that uses their data. Another advantage of binned scatterplots is that it is possible to control for the role of other covariates (Cattaneo et al. 2024). Figure 9.7 compares a traditional scatterplot (left) with a binned scatterplot (right). The traditional scatterplot suggests that there might be a linear relationship between the two variables, although this is not firmly established. The binned scatterplot can more clearly show the main patterns in the data. Furthermore, the confidence bands (shaded areas on the binned scatterplot) have a role similar to that of a confidence

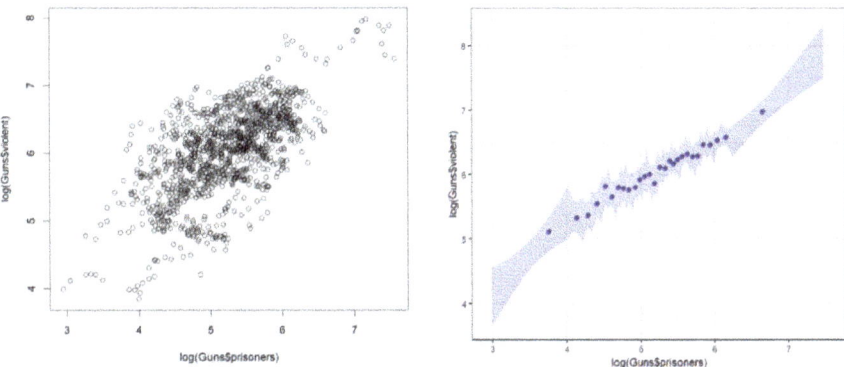

Fig. 9.7 Traditional scatterplot (*left*) and a binned scatterplot (*right*) (*Notes* Author's elaboration. See R code file for details)

interval for a function (Cattaneo et al. 2024, p. 1502). In other words, it seems that the shaded area in Fig. 9.7 (right) could accommodate a straight line. Therefore we find support for a linear relationship between the variables, on the grounds that a straight line would fit within the shaded area of the confidence band.

9.4 Communicating Graphs

Graphs are an important means of communicating data science results. Furthermore, presenting results in a clear way helps to build trust with decision-makers. Kenett and Redman (2019, Chapter 7) provide guidelines for the presentation of graphical results, organized according to the following four stages.

First, the facts should be presented as directly and accurately as possible. This is especially true when the results are less favorable than could have been hoped. If the results are unexpected and go against the previously accepted wisdom, then this should be clearly stated. Graphs should be explained as clearly as possible, such as explaining what is on the horizontal and vertical axes, what the lines correspond to, what is the frequency of observation (for time series graphs), how the performance variable is measured, and so on. It is also worth checking that the audience is following what you are saying by leaving ample time for questions during your presentation.

Second, a comprehensive picture of the main results should be presented. Forgetting to mention an important fact could be perceived as a deliberate lie.

Third, any relevant contextual knowledge should be provided, including the sources of the data, as well as information regarding the quality of the data. If the data is of unknown quality, or dubious quality, this should be clearly stated, as well as discussing how this could possibly influence the results.

Fourth, the analysis should be summarized, including a transparent discussion of the limitations of the analysis and possible alternative interpretations that could be driving the results.

9.5 Designing Dashboards

Dashboards in cars are familiar displays of useful information regarding key control metrics, such as gauges and dials to show the fuel level, speed, engine temperature, and so on. Data dashboards in corporate settings can be great ways to quickly and efficiently provide data and insights to decision makers such as top executives. Dashboards should be designed according to the above-mentioned principles of data visualization. Too many dashboards provide information that is obvious and unnecessary, while also presenting the information in a way that does not facilitate the rapid understanding of important numbers and trends. Some dashboards can be intimidating for users who lack advanced analytical skills, because it might not be clear how the information should guide decision-making. It is important not to overwhelm the user with lots of data but little information. Principles of pre-attentive processing can help guide the user's eye to quickly process the data and contribute to a clearer understanding of the circumstances.

Data dashboards can also provide users with the opportunity to interact with the information, for example if there are filters that allow users to select periods on which to focus, or to show disaggregated slices of performance data for different regions or product categories. Figure 9.8 shows an example of a corporate dashboard on fictional data, inspired by the principles in Wexler et al. (2017). Hovering with a mouse over datapoints in a chart can become a useful and efficient way of presenting details to users regarding the datapoints and an explanation on the data's sources. Dashboards should be designed such that users can dig deeper into the underlying details with just a few mouse-clicks. Comparisons are best represented with bar charts, and certainly not with pie charts. Line charts are useful for presenting trends and time series data. Wordclouds can be integrated into dashboards to highlight which keyword topics are more frequent in the textual analysis of complaints data. Vivid colours can be used to draw the user's attention to significant events or specific problems, but it could be counterproductive to use too many colours at the same time.

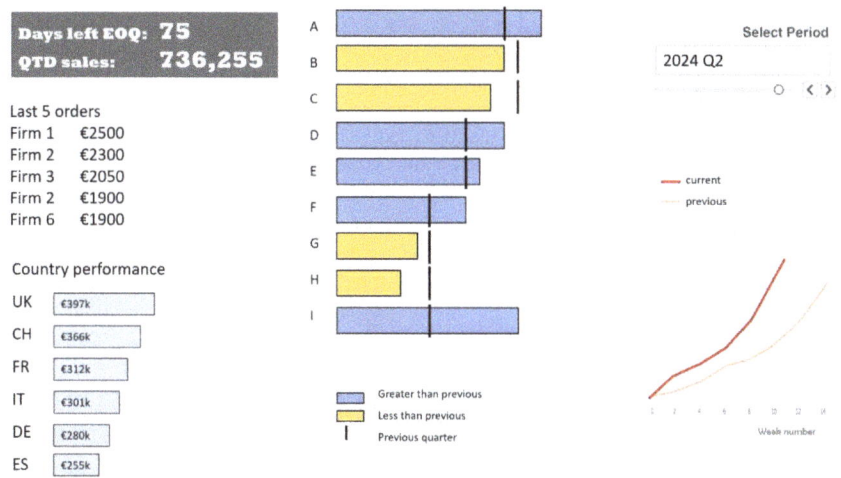

Fig. 9.8 Dashboard example made from fictional data (*Source* Author's elaboration, in the style of Wexler et al. [2017])

Further Reading

Tufte (2001) is a classic book on data visualization, and Healy (2018) is about data visualization specifically in R. Schwabish (2014) is an article-length discussion of data visualization for economists. Wexler et al. (2017) and Wells and Chiang (2017) focus on data visualization for dashboards. Wickham et al. (2016) presents the ggplot2 package for R; while Wickham et al. (2023)'s book on Data Science in R also covers data visualization. These latter two books are free to read online.

References

Cattaneo, M. D., Crump, R. K., Farrell, M. H., & Feng, Y. (2024). On binscatter. American Economic Review, 114(5), 1488–1514.
Healy, K. (2018). Data visualization: a practical introduction. Princeton University Press.
Kenett, R. S., & Redman, T. C. (2019). The real work of data science: turning data into information, better decisions, and stronger organizations. John Wiley & Sons.
Nussbaumer Knaflic, C. (2015). Storytelling with data—a data visualization guide for business professionals. John Wiley & Sons.
Schwabish, J. A. (2014). An economist's guide to visualizing data. Journal of Economic Perspectives 28 (1), 209–234.
Starr, E., & Goldfarb, B. (2020). Binned scatterplots: a Simple Tool to Make Research Easier and Better. Strategic Management Journal 41:2261–2274. https://doi.org/10.1002/smj.3199
Tufte, E. R. (2001). The Visual Display of Quantitative Information. Graphics Press; 2nd edition.
Wells, A. R., & Chiang, K. W. (2017). Monetizing Your Data: A Guide to Turning Data into Profit-Driving Strategies and Solutions. John Wiley & Sons.

Wexler, S., Shaffer, J., & Cotgreave, A. (2017). The big book of dashboards: visualizing your data using real-world business scenarios. John Wiley & Sons.

Wickham H., Çetinkaya-Rundel M., & Grolemund G. (2023). R for data science: import, tidy, transform, visualize, and model data (Second Edition). O'Reilly Media, Inc. Free to read online: https://r4ds.hadley.nz/

Wickham, H., Navarro D., & Pedersen T. L. (2016). ggplot2: elegant graphics for data analysis (3rd edition). New York: Springer. Free to read online: https://ggplot2-book.org/.

CART and Prediction

10

"Almost all of AI's recent progress is through one type [of AI], in which some input data (A) is used to quickly generate some simple response (B)," according to Andrew Ng, quoted in HBR (2023). In the case of OLS linear regression, seen in Chap. 5, the relation between the input data and the output variable is a simple linear relationship. The case of logistic regression is more complicated, because the relationship between inputs and outputs involves the logit link function, which is nonlinear. This chapter takes a tree approach, focusing on Classification and Regression Trees (CART). Instead of summarizing the relationship between the inputs and outputs in terms of a straight line (like for OLS) or a logistic curve (like for logistic regression), the relationship can be fitted by using a potentially unlimited number of splitting points to distinguish between datapoints according to their outcomes. This means that CART can fit the data much better, although the danger of overfitting becomes all the more serious, which has led to the development of robust techniques such as Random Forests.

Supplementary Information The online version contains supplementary material available at https://doi.org/10.1007/978-981-95-2433-4_10.

10.1 CART Models for Prediction

10.1.1 Introduction to CART

Classification and Regression Trees (CART) are a powerful set of techniques for making predictions. **Classification Trees** are for cases where the outcome is categorical, whether it be a binary variable (e.g. whether a particular email is spam or not) or a categorical variable (e.g. whether to go to work by bike, bus, train, or on foot, where the categories cannot easily be positioned along a unidimensional number line).[1] **Regression Trees** are for cases where the outcome is a continuous variable (such as salary, or price).

Consider the case of OLS linear regression with one explanatory variable x: $y = \alpha + \beta x + \varepsilon$. OLS tries to fit a straight line through the datapoints by varying the slope of the line β, and the position α where the line intersects the y axis. OLS is a parametric model, because it is restricted to cases where the only patterns it can find in the data are straight lines.[2] There are restrictions on how the input variable x can affect the response variable y: i.e. the relationship between x and y must take the form of a linear model. CART is a non-parametric technique that makes fewer assumptions about the relationship between x and y. For example, CART does not assume a linear relationship between x and y, and is also more flexible regarding the OLS assumptions about whether the error terms ε are normally-distributed, and whether the error terms ε are homoscedastic. (That said, an important requirement for CART is that the observations in the data should be statistically independent).

CART is more flexible than OLS in terms of finding a close fit to the data, and this flexibility makes CART far more vulnerable to the problem of overfitting. Overfitting has been likened to memorizing the idiosyncracies of particular datapoints, rather than focusing only on identifying the main trends (Provost and Fawcett 2013). To take an example, overfitting would be conceptually similar to memorizing the answers to a maths test before taking the test; as opposed to trying to understand the mathematical reasoning that will enable you to calculate the answers from the questions. After memorizing the answers to a particular test, you might get a perfect score when retaking the same test, but you would do very badly on a new test based on similar but different questions. The dangers of overfitting can be allayed for CART models by putting limits on the number of parameters used in the model (i.e. putting limits on the number of leaf nodes), by testing the CART model on new samples that were not used to train the initial model (such as cross-validation), and by using regularization techniques such as Random Forest.

---

[1] In other words, it is not possible to say that going to work by bike is "twice as much" as going to work by bus, which is "three times less" than going to work by train. These categories cannot be linked together in numerical terms.

[2] OLS can potentially investigate nonlinear relations, as seen below with a quadratic term. However, even in these cases, the fit to the data is restricted to taking a specific parametric functional form such as the quadratic curve, rather than having the flexibility to follow the patterns in the data.

10.1.2 Decision Trees and the Variable Space

Decision trees are a logical system for mapping the inputs to particular outcomes. Each node on the tree corresponds to a decision point where the observations are split one way or another. The decision tree is a set of hierarchically ordered steps that guide the input data to a conclusive state, which is represented by the final nodes on the tree (referred to as leaf nodes). Every node (except the root) has a parent node, and every node (except the leaf nodes) splits the data into two child nodes. The algorithm seeks to find the best ways of splitting the data such that the observations in each leaf node are similar, and that observations from different leaf nodes are different. The prediction for each leaf node (which corresponds to a predicted value for the outcome) is the average of the sample values for the outcomes from the observations in that leaf.

Figure 10.1 provides an illustrative example, from the topic of high-growth firms (HGFs), following Karlsson and Coad (2025). Observations are dropped into the tree (Fig. 10.1, left) at the root node at the top, and each decision node bounces the observation to the left or to the right. The splits in Fig. 10.1 (left) are dictated by the theoretical definition of an HGF. There are three leaf nodes, one of which corresponds to an HGF. Ultimately, the observation ends up in one of the leaf nodes, which contain the subsets that are defined by the data splits. Note that the splits in Fig. 10.1 (left) are dictated by our theoretical definition, not by the data, and as a consequence we expect that 100% of the observations in the HGF area of Fig. 10.1 (right) are HGFs, by definition. The first split, at the root node, is based on the requirement that HGFs should have at least 10 employees in the initial period. The second split, at the first (and only) internal node, is that HGFs should have an average annual growth rate of at least 20% over the three-year period. Observations that are "true" for both of these decision nodes end up in leaf node #2, which corresponds to HGFS; otherwise they are classified as non-HGFs and appear in leaf nodes #1 or #3.

Figure 10.1 (right) shows the same data as in Fig. 10 (left), but presented in a different way: in terms of a partitioning of the input space according to the tree's decision rules

10.1.3 CART Algorithm

Consider that we start with a set of data $\{x_i, y_i\}_{i=1}^{n}$, where x_i corresponds to the multiple input variables (features, or explanatory variables) and y_i corresponds to the output variable, for observation i. The CART algorithm will find the specific variable, and the specific value of that variable, that offers the single best split for the data, such that the child sets on the left and the right are as homogeneous (with respect to the outcome variable y) as possible within the child sets. In other words, the two child sets will be very different from each other, but each child set taken on its own will be relatively homogeneous. In mathematical notation, the optimal split occurs at location x_{ij} (for variable x_j and observation i) such that the child set

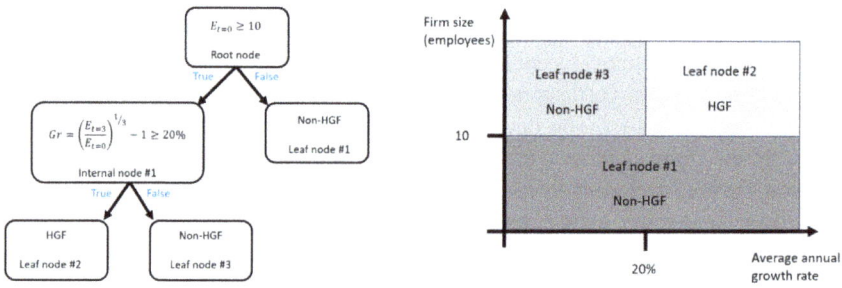

Fig. 10.1 The Eurostat-OECD (2007) HGF definition, represented as a decision tree (left) and represented as a partitioned input space (right) (*Notes* Author's elaboration, drawing on Karlsson and Coad [2025])

on the left $\{x_k, y_k : x_{kj} \leq x_{ij}\}$ and the child set on the right $\{x_k, y_k : x_{kj} > x_{ij}\}$ are as homogeneous as possible, according to a criteria such as minimizing the sum of squared errors for each child set taken individually, in the case of regression trees for continuous variables, i.e. minimizing this expression:

$$\sum_{k \in left} \left(y_k - \bar{y}_{left}\right)^2 + \sum_{k \in right} \left(y_k - \bar{y}_{right}\right)^2$$

In the case of classification trees that use categorical variables, we would not minimize the sum of squared errors, but would take a different objective such as minimizing the Gini impurity (Taddy et al. 2023).

Once the algorithm has found the optimal split, it takes the two child nodes and proceeds to split these two nodes further, using the same splitting procedure as mentioned above, to get four "grandchildren". The algorithm will stop splitting a parent node into two child nodes when instructed to do so by a **stopping rule**, which either could correspond to a limit on the leaf node size (e.g. stop splitting if the leaves have fewer than 12 observations), or a minimum improvement in the model fit (e.g. measured in terms of a minimum decrease in the loss function).

10.1.4 Greedy Algorithms

The CART algorithm is a greedy algorithm, which means that at each stage it seeks the best available split, but it is unable to think more than one step ahead. The algorithm chooses the largest immediate payoff, although this might not necessarily lead to the optimum. There is no consideration of whether an alternative split could lead to a better overall outcome in the long run.

Greedy algorithms can potentially get stuck at a local peak, and be unable to find the global peak. Louridas (2020, p. 62) uses the analogy of climbing a mountain:

10.2 R Example: Predictive Power of CART vs OLS, on Bivariate Data

The greedy heuristic would be to select the steepest path at each point (we assume that your climbing prowess is unparalleled). This will not necessarily lead you to the top: it may well lead you to a plateau, from which the only way is back. The real way to the top may lie through gentler slopes.

More generally, the CART algorithm is recursive, binary, and greedy. Its greedy nature has already been discussed. Its recursive nature comes from the fact that it starts with the "best" split, which splits the parent node into two child nodes, and then repeats the same splitting procedure on each of the child nodes. The binary label comes from the fact that each node is split into two child nodes.

10.2 R Example: Predictive Power of CART vs OLS, on Bivariate Data

This section focuses on applying CART and OLS to bivariate data with a relatively small number of observations. CART is not usually applied to small bivariate datasets, because it is a powerful technique designed for big data contexts. Nevertheless, focusing on a small bivariate dataset helps to clarify how CART differs from OLS, and how CART can sometimes give predictions that are far more accurate than those from OLS.

Consider the recorded speech data from the excellent book on Time Series analysis by Shumway and Stoffer (2025). The data are plotted in Fig. 10.2, with the speech indicator shown on the vertical axis, and the time period on the horizontal axis. There is a clear pattern in the data, such that the patterns within a pitch period tend to be repeated in subsequent periods. The figure highlights one example of a pitch period, and in the rest of this section we focus on the data within the highlighted pitch period.

We begin by focusing on fitting an OLS model to the subset of data (Fig. 10.3). The data is clearly non-linear, with a pattern that does not seem to fit well to a straight line. OLS regression is a parametric technique that tries to vary the intercept and the slope of a regression line such that the regression line finds the optimal fit to the datapoints. OLS regression has an advantage over CART in the sense that it is easier to interpret: the regression estimate $\hat{\beta} = -1.741$ implies that every time there is a unit change in the explanatory variable (here, time), we would expect a change of -1.741 in the outcome variable (here, the speech variable). Unfortunately, though, the OLS predictions are not very informative because the OLS line fits badly. In this case, the data are strongly non-linear, and there is no straight line that can fit well to the datapoints. The R code shows that the R-squared statistic is low: 0.012.

In a second OLS regression model, we add a quadratic term. A first remark is that, in OLS, we have to manually add the quadratic term ourselves if we want to allow for nonlinearities, whereas with CART, nonlinearities are automatically taken into consideration without having to ask. A second remark is that the

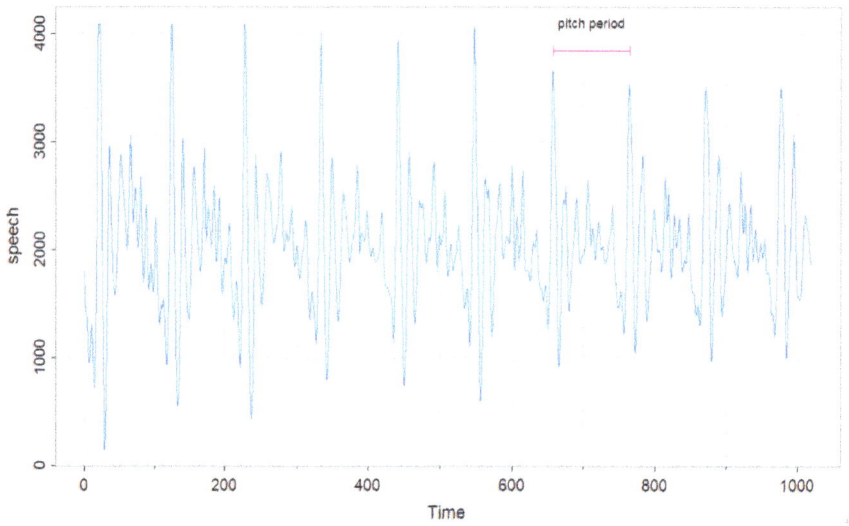

Fig. 10.2 Recorded speech data, using the data and R code from Shumway and Stoffer (2025)

quadratic term is not even significant (the t-statistic is 1.05) and the R-squared statistic has barely increased (from 0.012 to 0.022).

We now apply CART to the same data. CART automatically detects the nonlinear pattern in the data. This property of CART is referred to as **Automatic Interaction Detection (AID)**, and allows users to include nonlinear relationships and interaction terms in their model without specifying this in advance. With CART, instead of estimating a regression coefficient, we seek the decision nodes (or thresholds) that correspond to the optimal splitting rules for cutting the data into relatively homogenous subgroups. Figure 10.4 shows how the CART predictions (in red) are generally much closer to the actual datapoints than the OLS predictions, which are represented by the blue line of best fit. In this example, using the default settings, the CART model has cut the data up into 15 leaf nodes. CART would fit the data even better if we allowed it to take more than 15 leaf nodes.

CART does not give an R^2 statistic in the R output, in the same way that OLS linear regression does. Nevertheless, there are ways in which we can quantify the superior predictive power of CART over OLS in this simple example. First, we could look at the statistics on the models' residuals. For OLS, the median of the residuals is 22.85, and we can see this in the output of summary(lm(df$speech ~ df$time)). For CART, the median of the residuals is 2.40: much smaller than for OLS. Second, we could compare the predictive power by referring to the Mean Squared Error (MSE). The R code shows that the MSE for OLS linear regression is 247,916, whereas it is much smaller for CART, at 44,318.

10.2 R Example: Predictive Power of CART vs OLS, on Bivariate Data

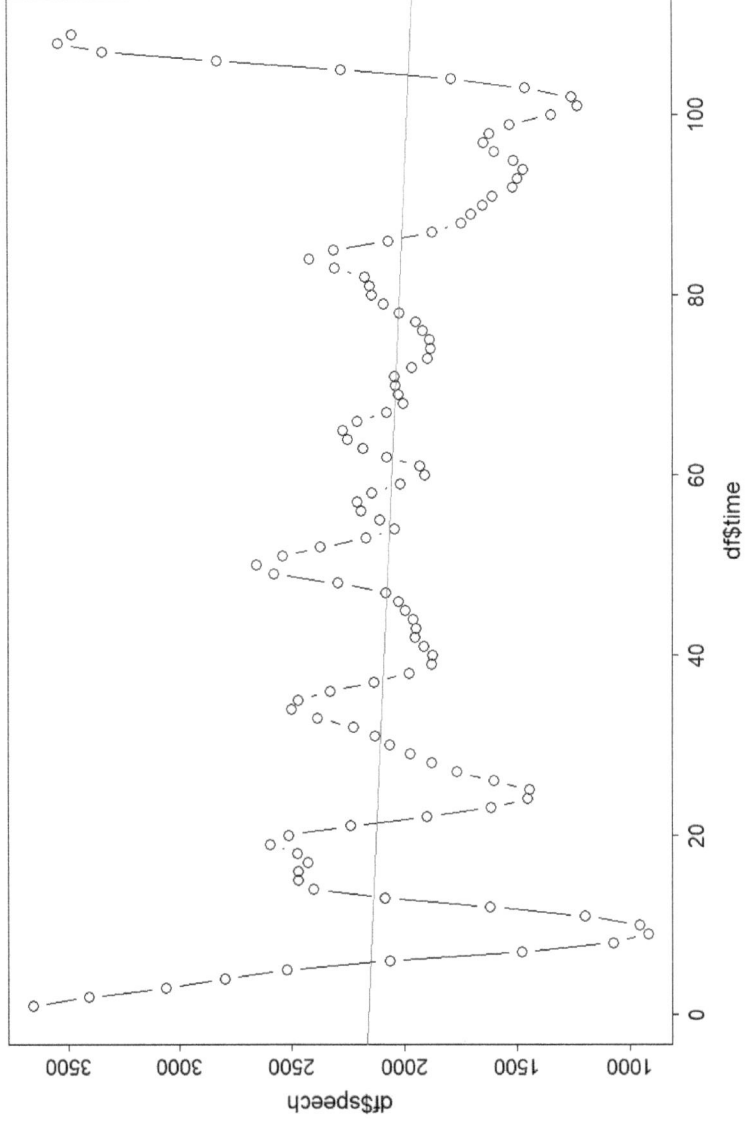

Fig. 10.3 OLS applied to a subset of the speech data (*Notes* Author's elaboration, see R code file for details)

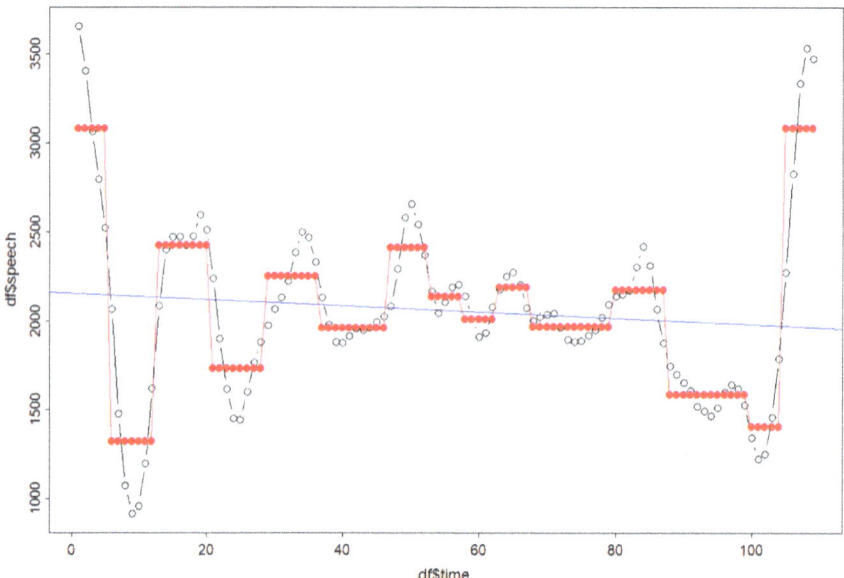

Fig. 10.4 OLS and CART predicted values (*Notes* Author's elaboration, see R code file for details)

10.3 Random Forests

Fitting a tree can lead to results that are unstable from one sample to the next. Furthermore, individual trees are susceptible to the problem of overfitting, which leads to poor out-of-sample predictive power. **Random Forests (RFs)** seek to address these problems by growing a large group of trees, and then obtaining predictions by averaging over many possible trees. Instead of fitting each tree in the forest to the same data, each tree is fitted to a different randomly-chosen subsample of the full dataset. Then, the Random Forest predictions come from averaging over the individual trees. Each tree now only holds part of the puzzle to solve the classification task. RFs can achieve reliability by drawing on the results aggregated from a forest of trees that are diverse, although each tree taken individually is required to have a satisfactory predictive performance (Genuer and Poggi 2020).

Figure 10.5 provides some intuition into the Random Forest approach, again referring to the context of the prediction of High-Growth Firms (HGFs) that we saw in Sect. 10.1.2. Figure 10.5 contains a forest of 8 trees, 5 of which are in favour of classifying observation i as an HGF. The information used to grow each individual tree is randomized, leading to a "forest" of multiple randomized classification trees. The principle of majority ruling is used to predict the outcome for each observation.

In a simple version of the Random Forests algorithm, trees are fitted on random (with-replacement) subsamples of observations i from the dataset. A more

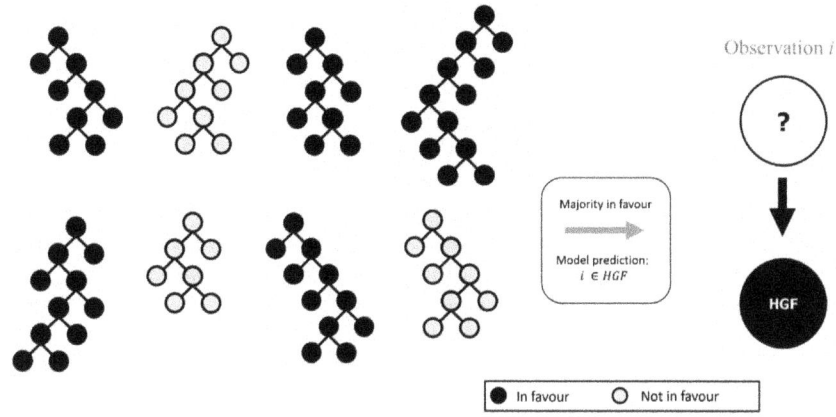

Fig. 10.5 How majority ruling in random forests leads to the classification of observations (*Notes* Inspired by Johan Karlsson, see Fig. 1.2 in Karlsson and Coad [2025])

advanced approach would be to fit trees not only to random subsamples of data, but also samples of explanatory variables, where each tree uses a random subsample of the available explanatory variables.

10.4 R Example: Lasso, CART and Random Forests on Real Estate Data

This section features an application of CART and random forests on real estate transactions for residential lots in the Tokyo area, using data for the 1st-4th quarters of 2024. Data are free to download from the website of MLIT: https://www.reinfolib.mlit.go.jp/realEstatePrices/.

10.4.1 Cleaning the Data

In the R code file, we begin by taking a look at the data. It clearly needs tidying up. We give clearer names to the columns to clarify what the variables are. To simplify the data, we remove the variable "Tokyo" because all observations in the sample have the same value for this variable.

Next, we set up the dependent variable. When we type `colSums(is.na(mlit))`, we can see that `price_m2` has lots of missing values. Fortunately, this is not a serious problem, because we have more complete data for the variables price and m2, and can construct a new indicator of price per m2 ourselves. This is done using `mlit$price_m2_calc <- mlit$price / mlit$area`. Then we check whether our newly-calculated variable (mlit$price_m2_calc) corresponds to the original variable (mlit$price_m2), for the cases where the original variable (mlit$price_m2) is non-missing. Scatterplots show that mlit$price_m2 and mlit$price_m2_calc are highly correlated, with datapoints almost perfectly positioned on the diagonal line. The correlation is extremely high at 0.999, for the cases that we can observe. We will assume that mlit$price_m2_calc is a satisfactory indicator for our purposes. We can now delete mlit$price_m2 because we use our own calculated variable instead.

Now, we check the explanatory variables. Some variables need to become numeric variables, such as distance to the nearest train station, mlit$statn_dist. We also create a variable statn_dist_sq, which could be important if there are nonlinear effects (in our Lasso regression), such that distance from the station is considered to be extremely inconvenient if it is above a certain threshold.

The variable "floor_area" is numeric except for one case which takes the value "2,000 ·u or greater." We replace these cases with a value of 2000 and convert the variable to numeric.[3]

Regarding the other explanatory variables, we convert the categorical variables to being factor variables.

In the case of missing values, we convert NAs into an extra factor level using the function addNA. This allows us to avoid dropping these instances of what would otherwise be missing values. If our empirical analysis were to indicate that the factor level corresponding to NA were an important predictor, we could explore this in more detail.

In other cases of missing values, we delete some rows in our database if there are missing values for variables that we think are important enough to keep in the main analytical model rather than omitting: statn_dist (and hence statn_dist_sq), floor_area, and cons_year.

More generally, we remove any remaining missing values using mlit <- na.omit(mlit), which reduces the dimensions of our data matrix from $11{,}225 \times 27$ to $10{,}705 \times 27$.

10.4.2 Lasso

Now that we have cleaned the data, and before applying CART, it seems worth fitting a quick Lasso. To prepare the data for the gamlr function, we put the data into sparse model matrix format. The Lasso path plot is shown in Fig. 10.6. In the unconstrained OLS model, there would be over 2000 coefficients, but in the AICc-selected Lasso model, there are 1271 coefficients.

---

[3] Ideally, we would replace the case of "2000 or greater" with the average value for this category, which would probably be much larger than 2000. However, we lack information on this category, and we simply replace it with 2000. If this variable turns out to be particularly important, it might be worth exploring in more detail regarding the best way to replace this open-ended residual category.

10.4 R Example: Lasso, CART and Random Forests on Real Estate Data

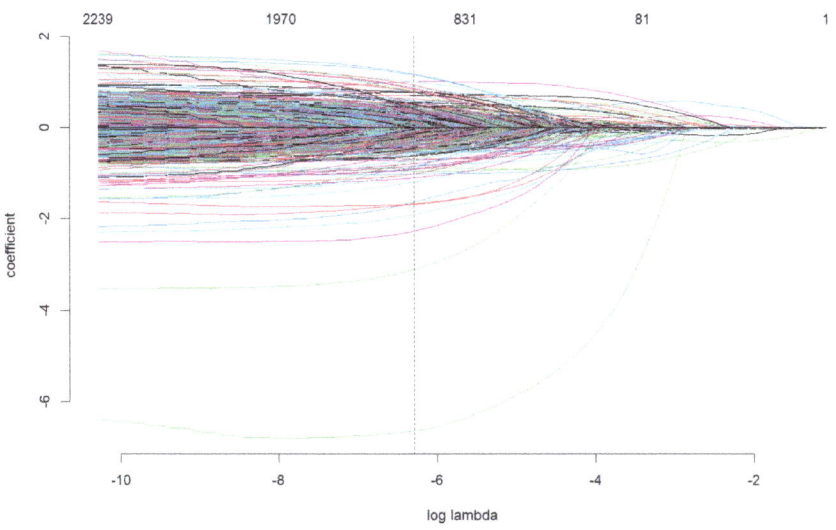

Fig. 10.6 Lasso path plot (*Notes* Author's elaboration, see R code file for details)

Lower values of the dependent variable (log of price per square metre) are strongly negatively associated with certain districts (Higashishimbashi, Nishiwake-cho, and Tomodamachi), and with the land use corresponding to the category of "Office, Warehouse, Parking Lot, Shop". Among the most positive Lasso coefficients, there are variables corresponding to ward (Chiyoda), district (Uenokoen) and station (Aoyama Itchome), as well as properties with a high value for Floor Area Ratio (land upon which tall buildings and skyscrapers are allowed).

Focusing on some specific variables, station distance is negatively associated with the outcome (log of price per square metre), but that the construction year of the property and the floor area are positively associated with the outcome variable.

10.4.3 CART

We now fit trees to the MLIT data, starting with a single explanatory variable, which is the distance from the station. Figure 10.7 (left) shows the corresponding dendrogram. Vertical lines on the dendrogram have a length that is proportional to the reduction in heterogeneity that results from the splitting, which is an indicator of the improvement in model fit that is due to the partitioning. As such, the first split in Fig. 10.7 (left) seems to be more important than subsequent splits. For the splitting rule "`milt$statn_dist < 4.5`", if the condition is true we get the child node on the left (14.00), and if the condition is false we get the child node on the right (13.52). Figure 10.7 (left) shows that the log of transaction prices/m^2 are the highest for properties less than 4.5 min from the nearest station, but

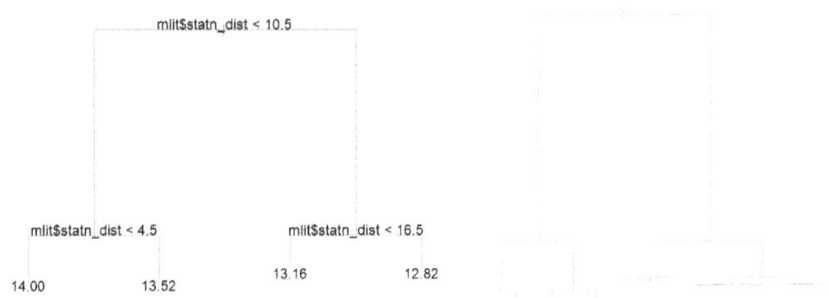

Fig. 10.7 Trees fitted to the MLIT data. *Left*: univariate tree. *Right*: univariate tree that is a ridiculous case of overfitting (*Notes* Author's elaboration, see R code file for details)

they are the lowest for properties greater than 16.5 min from the nearest station. Figure 10.7 (right) shows a ridiculous example of overfitting. The bottom of the dendrogram shows a large number of splits but with a negligible improvement in model fit (because the vertical lines of the dendrogram associated with these lower splits is short). If the text is plotted on the dendrogram, it is hard to read. Figure 10.7 (right) is surely affected by the problem of overfitting.

Figure 10.8 presents a more complex tree that has more features. The first split, at the root note, refers to whether the station distance is less than 10.5 min. Other important variables seem to be the floor area and the year of construction. The leaf node corresponding to the highest values of the dependent variable (log of price per m^2 = 14.77) is for cases that are less than 10.5 min distance from a station, with floor area greater than 232.5, and a construction year that is more recent than 2017. In contrast, the leaf node corresponding to the lowest outcomes (log of price per m^2 = 12.08 on average for that leaf node) is for properties with a distance of greater than 20.5 min from the nearest station, that were constructed in 1999 or earlier.

10.4.4 Random Forests

We end this example by fitting a random forest to the MLIT data, using the ranger package. We include a fairly large set of features (explanatory variables).

Random forests are more robust to the problem of overfitting than single trees. However, a drawback of random forests is that the results are not as easy to interpret as individual trees (such as the example in Fig. 10.8). Nevertheless, random forests analysis can include an analysis of variable importance, which is carried out by measuring the predictive performance of individual trees in the forest that are formed by taking into account different subsets of variables. Put differently, the importance of a variable can be calculated in terms of the increase in error that

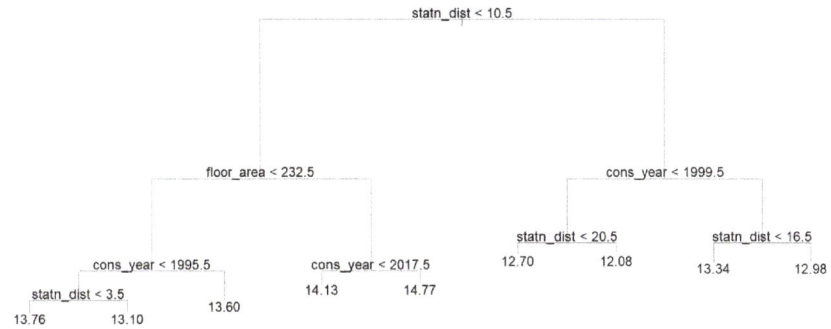

Fig. 10.8 A more complete tree, fitted to the real estate data (*Notes* Author's elaboration. See R code file for details)

```
> sort(RFmlit$variable.importance, decreasing=TRUE)
    cons_year      cov_ratio    statn_dist_sq       roadtype      floor_area      statn_dist       area_type        planning
    754.81782      691.44504         483.52253      448.47885      434.36976       370.40484       367.88713       242.49019
    structure          width             shape      direction        purpose         quarter              TF
    207.21372      201.36022          77.82390       52.66751       42.13180        31.56010        23.01202
```

Fig. 10.9 Results of the Random Forest analysis of variable importance (*Notes* Author's elaboration. See R code file for details)

arises when a particular variable is not used to define the tree splits (Taddy et al. 2023, p. 273; see also James et al., 2001, Section 8.3.3).

Regarding the options for the ranger command, we fit a forest with 100 trees and a minimum node size of 50. The option `importance = "impurity"` refers to the calculation of variable importance.

Figure 10.9 shows that the following variables are identified by the random forest procedure as being the most important: year of construction, coverage ratio, distance from the nearest station squared, road type, and floor area.

Further Reading

James et al. (2021) and Taddy et al. (2023) contain excellent chapters on trees and CART in R. Genuer and Poggi (2020) is a book on Random Forests focusing specifically on R.

References

Eurostat - OECD (2007). Eurostat-OECD Manual on Business Demography Statistics. Office for Official Publications of the European Communities: Luxembourg.
Genuer, R., & Poggi, J. M. (2020). Random forests in R. Springer International Publishing.

HBR (2023). HBR Guide to AI Basics for Managers. Harvard Business Review Press. Massachusetts: USA.

James, G., Witten, D., Hastie, T., & Tibshirani, R. (2021). An Introduction to Statistical Learning with Applications in R. 2nd Edition. Springer Nature, New York, NY, USA. Free to download from: https://www.statlearning.com/

Karlsson, J., & Coad, A. (2025). Searching for Gazelles in the Forest: The Potential of Random Forests to Identify High-Growth Firms. Chapter in: Anna Ujwary-Gil & Anna Florek-Paszkowska (Editors); AI, Analytics and Strategic Decision-Making. New York, USA: Routledge. https://doi.org/10.4324/9781003507840-3.

Louridas, P. (2020). Algorithms. MIT Press Essential Knowledge Series. MIT Press. Cambridge MA, USA.

Provost, F., & Fawcett, T. (2013). Data Science for Business: What you need to know about data mining and data-analytic thinking. O'Reilly Media, Inc.

Shumway, R. H., & Stoffer, D. S. (2025). Time Series Analysis and Its Applications: With R Examples. Fifth Edition. Springer Nature.

Taddy M., Hendrix L., & Harding M.C. (2023). Modern Business Analytics. Practical Data Science for Decision Making. McGraw Hill, New York, NY.

Text as Data 11

Text data is emerging as an exciting new data source. As the costs of storage drop, and as textual records move to digital platforms, there are growing possibilities for text as data. Text data has been used to obtain new insights in various areas, from finance (predicting asset price movements) to macroeconomics (forecasting variables such as inflation and unemployment) to marketing (to study customer decisions) to political science (studying the dynamics of political debates). In the area of business data science, there are a variety of communication channels that could feed into quantitative analysis of text, such as customer support conversations, product descriptions, product reviews, news, comments, complaints data, blogs, tweets, and anonymized employee emails. Text data is basically another way for firms to learn from customers.

It may be counterintuitive to some that text data can be the subject of quantitative analysis. Text data is what data scientists would call unstructured data: it does not come in the form of a table of numbers, although we can convert text into a table of numbers. Essentially, text is just another form of data. Text data is a rich complement to more structured variables in traditional databases. Text data can be used for tasks such as monitoring (e.g. content moderation to suppress hate speech), prediction (e.g. for stock market trading) and causal inference (e.g. for studying which ads will cause the largest increase in sales; Grimmer et al. 2022). Text requires a lot of pre-processing, however, before it can be used in quantitative analysis.

Supplementary Information The online version contains supplementary material available at https://doi.org/10.1007/978-981-95-2433-4_11.

This chapter seeks to familiarize the reader with the basics of quantitative analysis of text data. We start with the "bag of words" model, which is usually considered to be the starting point for the analysis of text data, because it is a relatively simple model. Despite its many limitations, the bag of words approach works quite well in many contexts. We then discuss topic models, which introduce an intermediate layer of unobserved topics into textual analysis to attempt to better understand the meaning in texts. The chapter also contains some examples in R.

11.1 The Bag of Words Model

The bag of words is a simple and well-known model for text representation. A rough analogy could be that a piece of paper with text printed on it is cut up with scissors into individual words, which are then jumbled together in a bag. A document is therefore represented as the set of words that it contains, and by counting up the number of times that particular words appear. The bag of words model achieves its simplicity by discarding a lot of potentially information (on word order, context, synonyms, punctuation, and so on). It is a parsimonious representation that is still surprisingly useful and is often used as a first stage of text analysis before potentially moving on to more complicated models (such as topic models).

Despite its simplicity, there are a number of methodological choices that go into the bag of words representation, that are discussed in the next subsections. The tokenization process takes the initial text data and cuts it up into individual tokens, while also reducing complexity by deleting punctuation, converting all letters to lower case, and filtering out words that are either extremely common or extremely rare. The text data can thereby be vectorized for each document, and then converted into a numerical matrix, the Document-Term Matrix (DTM).

11.1.1 Tokenization

The bag of words approach considers that a text document can be split up into a number of tokens. This process is called tokenization. The raw text is cut up into individual words, and thereby ignores the complex structure of language that comes from grammar, sequences of words, and subtle forms of dependence among the words in documents. In English, tokenization is relatively easy because individual words are separated by white space. In other languages, such as Japanese and Chinese, words are not separated using white spaces, and so a different approach is required (Hvitfeldt and Silge 2022).

Tokenization involves a number of steps in order to reduce the complexity of the raw text. Case normalization converts words into lower case in order to standardize the tokens. Punctuation is usually removed, which in many cases is fine, but could be problematic in cases of character strings such as ¯_(ツ)_/¯ that might appear in social media messages.

In English, the unit of analysis is usually considered to be a word, although this need not always be the case. At one extreme, it is possible to imagine splitting up a chunk of text into the 26 individual letters of the English alphabet. This could be useful in terms of yielding a vocabulary size of 26, but obviously the approach would be flawed because an individual letter conveys far less meaning than when letters are grouped together into words. On the other hand, it could be possible to split up text into longer chunks such as sentences. This would be problematic from the perspective of text analysis, because it is unlikely that exactly the same sentence will appear more than once in a given corpus of documents. Focusing on the word level is a compromise between these two extremes (Grimmer et al. 2022). However, in some cases it is useful to take combinations of words ("**ngrams**") such as **bigrams** (a unit comprised of two consecutive words) or **trigrams** (three consecutive words). Ngrams are easy to create (as simple as changing the options in R) and require no expert linguistic knowledge.

Consider the following text:

National Security regulations implemented at the White House

Some bigrams are useful, because the meaning of the bigram is distinct from the meaning of the words taken individually: national_security, white_house. Other bigrams are less useful, however, because they merely correspond to two different words haphazardly stuck together, such as regulations_implemented. Tokenization in terms of bigrams or trigrams will therefore work better if it is combined with a technique for automatic variable selection (e.g. Lasso regression) such that a larger number of irrelevant bigrams and trigrams can be removed from the regression model.

Ngrams considerably increase the size of the vocabulary, because the addition of word-pairs to the vocabulary dimension can vastly increase the number of columns in your data matrix (e.g. the document-term matrix). Relatedly, focusing on bigrams will considerably increase the number of explanatory variables in your text regression. As such, the tokenization process for bigrams takes more time, the data matrix requires more memory, and the computations will require more processing power, and therefore a focus on bigrams and trigrams is rarely worthwhile. An alternative to bigrams would be Named Entity Extraction, which involves manually pre-processing the text data to identify certain bigrams or trigrams with specific meanings (such as creating a token for "Silicon Valley"). This procedure requires expert knowledge and takes time for handcoding for each particular case (Provost and Fawcett 2013).

11.1.2 Stopwords

Pruning of "stopwords" refers to the removal of extremely common words, that are used as a sort of "glue" to hold sentences together, but have little meaning when considered as individual tokens. Examples of stopwords include words such as the following: the, and, if, but, who, what, the, they, their, a, or. Numbers are also often removed as part of the process of tokenization and the removal of stopwords. In practice, there are standardized list of stopwords that are often integrated into text analysis software. In a first stage, the pruning of stopwords could rely on a generic list of stopwords. There may also be a second stage, where (depending on your particular application) the corpus of documents being analyzed has an unusually large number of certain words, which have a similar role as stopwords, although they might not appear in general-purpose lists of stopwords. This could be the case of a corpus of documents that all focus specifically on the topic of hedgehogs. As such, there may be a second stage of pruning that involves an application-specific stop word list (such as removing the word "hedgehog").

One researcher's stopword could potentially be another researcher's key term. There are many exceptions. Stopwords such as "the" and "who" could be problematic in the situation where references to music groups such as "The Who" and "The The" might completely disappear as part of the process of normalizing to lower case, and then eliminating stopwords.

One well-known early contribution to text analysis is the study of the authorship of the Federalist papers by Mosteller and Wallace (1963). The Federalist papers are a corpus of 85 documents written by Alexander Hamilton, John Jay, and James Madison in the late 1700s, written to support the US constitution. Authorship of 73 out of the 85 documents is firmly established. However, because different sources of information were in conflict, it was not clear who wrote the other 12 documents. Mosteller and Wallace therefore learnt about the writing styles of the authors by studying the frequency of "filler words" used the clearly-labelled 73 documents, and applied these insights to the remaining 12 documents (which had not yet been "labelled" with an author). In particular, the use of stopwords (such as "upon") helped to identify authorship. This story helps to emphasize an important point for text analysis in general: methodological choices should be consistent with the purpose of your study.

In addition to the pruning of overly common stopwords, it is recommended to remove exceptionally rare words. For example, Taddy (2019) recommends removing words that occur in fewer than 15% of documents. The exclusion of rare words is considered to be a "necessary evil". Words that occur only once are useless for comparing documents. For example, if you analyze the current book regarding the appearance of words such as Timbuktu, platypus, and nudiustertian, you will find that these three words occur exclusively in this chapter (Chapter 11, on "Text as data"), but these words are so rare that they will not help to distinguish between the meaning of this chapter and other chapters.

Fig. 11.1 gives an overview of an example in R, where raw text (the first 6 lines from the novel "A Tale of Two Cities" by Charles Dickens) undergoes tokenization (yielding a vector of 86 words) and then the removal of stopwords (yielding a vector of 22 words).

11.1 The Bag of Words Model

Fig. 11.1 Processing the raw text with tokenization and the removal of stopwords (*Notes* Author's elaboration, details in R code file)

11.1.3 Stemming

Stemming is another processing step for reducing the complexity of text data that seeks to improve the reliability of the counting of tokens in a bag of words model. If a document contains variations on a particular word, then it could be preferable to cut words to the root, discarding common suffixes such as –s and –ing, and counting the number of appearances of the root. For example, the stem "analy" could be considered to be the root of several related words, such as analytics, analysis, analyze (American English spelling), analyse (British English spelling) and analyst. The Porter stemming algorithm (Porter 1980) is a well-known and widely used stemming tool. Stemming can be a helpful procedure, but stemming tools can sometimes be too aggressive, and mistakenly reduce unrelated words to a common root. Again, whether to apply stemming depends on the context of your research question.

11.1.4 Document Term Matrix (DTM)

Once the previous steps have been done (tokenization, pruning of stopwords, stemming, etc.) then the remaining tokens can be counted and represented in quantitative terms in a Document Term Matrix (DTM). The DTM is constructed according to the logic of the Bag of Words approach: it ignores the order of words, and the context in phrases, but nevertheless it can lead to some useful results. After the pre-processing that was described in the previous subsections, the DTM is the stage where we convert text data in to a matrix of numbers.

The DTM is an $N \times J$ matrix that consists of N rows and J columns. Each row corresponds to one if the N documents, indexed by i. Each column corresponds to a token (indexed by j), and therefore the number of columns corresponds to the total size of the vocabulary J. Note that, as you add more text data, the size of the vocabulary is also expected to grow. The DTM is high-dimensional data that is in the form of a sparse matrix, with most of the cells containing zeroes. Each cell W_{ij} in the matrix corresponds to the number of times (i.e. a non-negative integer count variable) that each token j appears in a document i.

Fig. 11.2 shows 2 **documents** that constitute our corpus. A **corpus** is a collection of documents (the plural would be "corpora"). A document could be anything from a single sentence to a large book. In Fig. 11.2, the "documents" are short jokes in the form of a sentence. We have 2 rows, one each for Document1 and Document2. We have 6 columns, because the vocabulary size (after tokenization and removing stopwords) is 6. (Arguably, the removal of stopwords in this example is quite severe, because we only have 6 tokens left in both documents.) Only one of the tokens in the vocabulary ("knock") appears in both documents. Representing text data in DTM format clearly discards a lot of useful information from the raw data. The jokes in Documents 1 and 2 were not very funny to start with, but nobody would laugh at a joke when it is represented in DTM format, because so much of the meaning has been lost. Nevertheless, in practice, quantitative analysis on DTMs can often lead to valuable new insights.

DOCUMENT 1:
Knock knock. Who's there? Boo. Boo who? Don't cry, it's just a joke!
DOCUMENT 2:
Knock knock. Who's there? Cow says. Cow says who? Cow says moo, not who!

After tokenization and removing stop words, we get this DTM:

	A	B	C	D	E	F	G	H
1		knock	boo	cry	joke	cow	moo	row total
2	Document 1	2	2	1	1	0	0	6
3	Document 2	2	0	0	0	3	1	6
4	column total	4	2	1	1	3	1	

Fig. 11.2 From raw text to the corresponding representation in a Document-Term Matrix (DTM) (*Source* Author's elaboration)

11.1.5 Term Frequency and tf-idf, with an R Example

The tf-idf is a useful and well-known indicator of how important word j is for document i. tf-idf stands for Term Frequency—Inverse Document Frequency. The tf-idf is formed by combining two elements: tf and idf. The equation for term frequency (tf) is a simple count of the number of times that word j appears in document i:

$$tf_j = W_{ij}$$

Words that appear frequently in one document are important for that document, but what about exclusivity? Do these words also appear a lot in other documents? The idf term takes into account the exclusivity of a word in a document. The idf term is essentially a penalty for frequent words that appear in all documents in the corpus.

The idf is commonly measured as follows (Silge and Robinson 2017; Grimmer et al. 2022)[1]:

$$idf_j = log\left(\frac{Total\ number\ of\ documents}{Number\ of\ documents\ containing\ j}\right)$$

For word j in document i. According to the idf, rare terms are special and have high scores. Stopwords, however, will have an idf that is low. Putting these two together, we get the tf-idf equation:

$$W_{ij}^{tf-idf} = W_{ij} \times log\frac{N}{n_j}$$

where N is the number of documents in the corpus, and n_j is the number of documents that contain the word j. Note that the tf part is document-specific: it refers to a single document i; whereas the idf part is calculated for all documents in the corpus.

The R code file shows an example of the tf-idf statistic. In this example, we analyze a corpus of three documents which are classic texts in Economics: The Essays of Adam Smith; the Communist Manifesto by Karl Marx and Friedrich Engels; and David Ricardo's "On The Principles of Political Economy, and Taxation."

The R code starts by loading these books into R using the `gutenbergr` package. The next step is tokenization, after which the `ggplot2` package is used to create a horizontal bar chart of the most common tokens. As a simple preliminary step in data exploration, Fig. 11.3 shows the wordclouds for the three books.

The next part calculates the tf-idf statistics for the three books. We calculate a matrix "`book_words`" that contains 4 columns: word, n (number of appearances

---

[1] See also Provost and Fawcett (2013, Chapter 10) for a variation which includes a constant term (equal to 1) in their equation.

Fig. 11.3 Wordclouds: Adam Smith (*left*), Karl Marx (*centre*), and David Ricardo (*right*) (*Notes* Author's elaboration, see R code file for details)

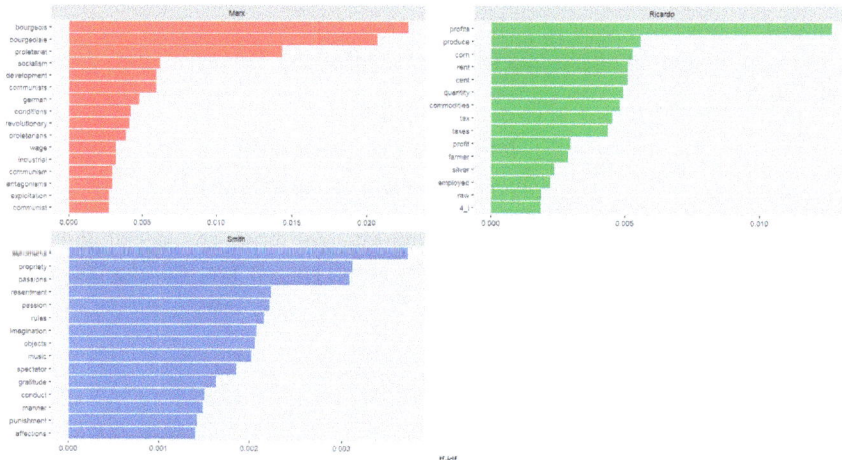

Fig. 11.4 tf-idf statistics for Smith, Marx, and Ricardo (*Notes* Author's elaboration, see R code file for details)

of the word), book, and total (total number of words in the book). We then apply the function "`bind_tf_idf()`" to the matrix "`book_words`" and plot the tf-idf statistics for the three books (as shown in Fig. 11.4).

11.1.6 Text Regression

Text regression can fit well with the Bag of Words logic. Lasso regression, in particular, works well in such cases of high dimensional data, because it has a rigorous approach to handling overfitting, as well as reducing the complexity of the regression model by selecting which unimportant variables can be discarded. Examples in R for applying Lasso regression on text data can be found in Hvitfeldt and Silge (2022) and Taddy et al. (2023). Taddy et al. (2023) apply Lasso to a corpus of 60,000 online reviews of various businesses (such as restaurants) written by customers. These online reviews are also "labelled" with an overall score that ranges from one star to five stars (in the spirit of "supervised learning"). As such, text regression takes the form of predicting the dependent variable (the number of stars awarded) on the basis of the text data as represented by the DTM. Text regression is less useful in other cases, however, such as where there is no clear choice of what could be the dependent variable, and the goal instead is to discover natural groupings in the data, such as how documents are associated to particular topics (in the spirit of "unsupervised learning").

11.2 Sentiment Analysis, with R Example

Sentiment data can be useful in business applications, for example to see the sentiment of customer social media data, customer feedback, complaints data, etc. This subsection discusses sentiment analysis in R, and continues with the Charles Dickens example of Sect. 11.1.2 to analyze the dynamics of emotions across the novel.

Words carry an emotional meaning. As such, text analysis can be extended to take into account the sentimental value of words and phrases. To this end, sentiment dictionaries have been developed to give an emotional label to words. The counting of word tokens (according to the bag of words approach) can be combined with a way of labelling words in terms of sentiment. In this way, the total sentiment of a document (such as a novel) can be calculated by summing up the sentiment scores for the words in the document's text. This section looks at one way (among many ways) of analyzing the sentimental dimension of text, taking a similar approach to Silge and Robinson (2017, Chapter 2).[2]

The `tidytext` package in R offers a number of sentiment lexicons that are based on unigrams. In particular, we will focus on the following three. `AFINN`, from Finn Årup Nielsen,[3] gives words a score that ranges from −5 (for strong negative sentiment) to +5 (for strong positive sentiment). The lexicon `bing`, from Bing Liu and colleagues,[4] sorts words into positive and negative categories in a binary way. Finally, `nrc` puts words into a number of binary categories, such as positive emotion, negative emotion, anger, fear, joy, and sadness. Of course, many words in English are relatively neutral, as far as emotions are concerned, and so they would not be given.

The R code file begins by using "`get_sentiments()`" to load up the three sentiment lexicons that we will be using (`AFINN`, `bing`, and `nrc`). We remove the first 75 lines of the Dickens novel, because they correspond to preliminary material and a table of contents. We then convert the "`dickens`" object into a format that

---

[2] Need it be reminded, Silge and Robinson (2017) is free to read online, at https://www.tidytextmining.com (last accessed 23 July 2025).
[3] https://www2.imm.dtu.dk/pubdb/pubs/6010-full.html (last accessed 23 July 2025).
[4] https://www.cs.uic.edu/~liub/FBS/sentiment-analysis.html (last accessed 23 July 2025).

is compatible with the sentiment lexicons. In particular, the sentiment lexicons have a column called "word", and if we set up a column called "word" in our data file, we can join the two together using the "inner_join()" command. The R code plots a horizontal bar chart that depicts the 10 most common negative and positive words (Fig. 11.5).

The R code also plots a wordcloud (Fig. 11.6) that contrasts the most frequently-occurring negative sentiment words, with the most frequently-occurring positive sentiment words.

Finally, we follow a procedure in Silge and Robinson (2017) that presents a sophisticated way of tracking the evolution of sentiment over the course of the novel. In the R code file, the novel is cut up into 80-line sections, and for each section we calculate the net sentiment score (the difference between positive and negative sentiment). A useful operator for this task in R is the integer division operator, "%/%". We can bind all the sentiment scores for the 80-line sections and put them together in a plot (see Fig. 11.7).

Some comments can be made on Fig. 11.7. Similar patterns are observed for the three sentiment lexicons. The central panel, corresponding to Bing et al., seems rather erratic in places, and overall the sentiment seems to be quite negative. NRC,

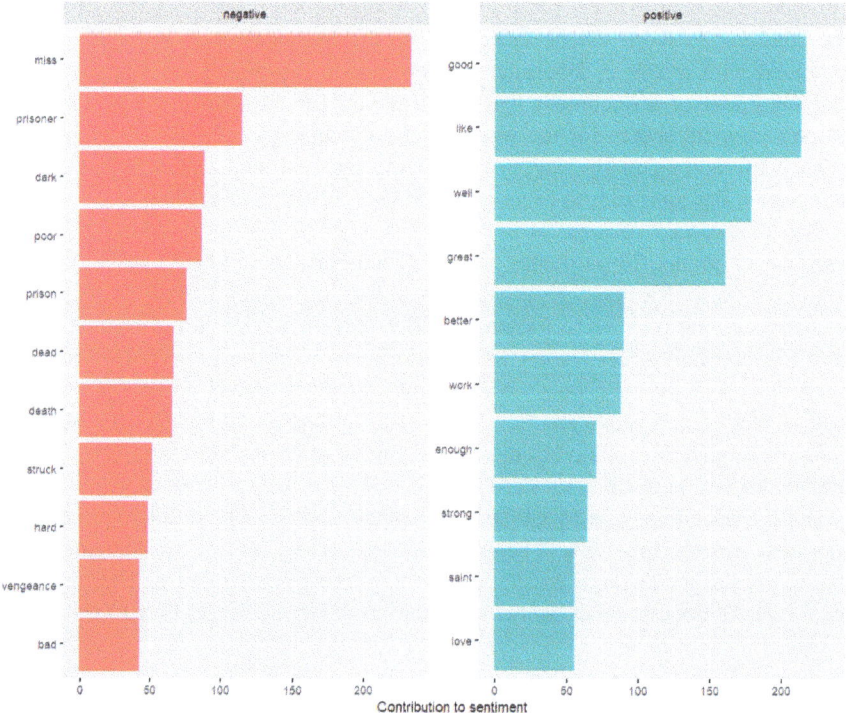

Fig. 11.5 The most frequent negative-sentiment and positive-sentiment words in the novel "A Tale of Two Cities" by Charles Dickens (*Notes* Author's elaboration, see R code file for details)

Fig. 11.6 Wordcloud for the most negative and positive words (in terms of sentiment) in the novel "A Tale of Two Cities" by Charles Dickens (*Notes* Author's elaboration, see R code file for details)

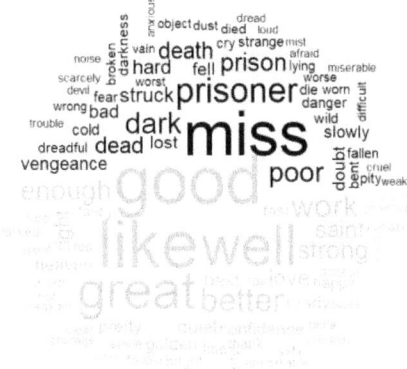

in contrast, seems to be more positive in terms of sentiment score. AFINN seems to be in between the two other sentiment lexicons. Overall, the emotional dynamics in the novel seem to indicate that the second half contains more negative sentiment than the first half, although in the last section of text there does seem to be a happy ending.

11.3 Topic Modelling

According to the bag of words method, the analysis of text data involves dealing with high-dimensional data (such as high-dimensional DTMs). However, topic modelling takes a different approach compared to bag of words. Topic modelling adds an intermediate layer of topics between the individual words and the document. These topics are not observed, but they are inferred from the data. Topic modelling treats each document as if it were a mixture of topics, and each topic is considered to be a mixture of words. Documents are allowed to overlap with each other in terms of topics, because the same topics can be used to help characterize multiple documents. Put differently, there is no attempt to attach one document to only one topic, or one topic to one document.

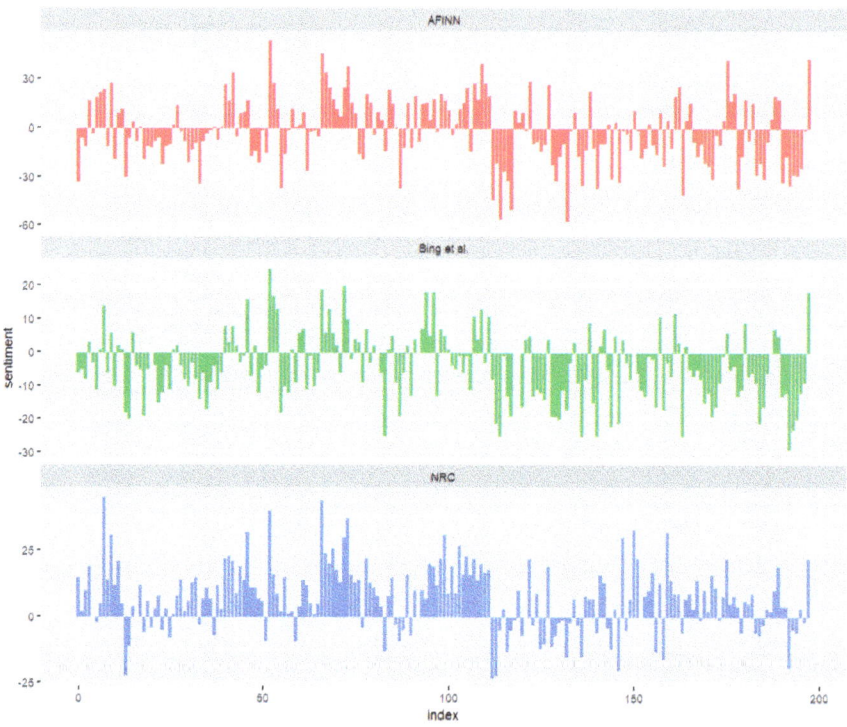

Fig. 11.7 Sentiment analysis for Dickens, for 3 sentiment dictionaries (*Notes* Author's elaboration, see R code file for details)

Figure 11.8 shows how topic models can be thought of as adding an intermediate layer between the words and the document (Provost and Fawcett 2013). Figure 11.8 shows that the words "purr", "dog" and "furry" are associated with the topic "pets." Note that the words "pets" and "children" do not actually appear in the text, hence a bag-of-words approach would not find any terms corresponding to "pets" or "children" in the text passage, even though the passage is quite clearly focusing on the themes of pets and children. With topic modelling, the topics are learned from the data, using unsupervised data mining to explore the data and discover which topics it might contain.

Documents are then classified on the basis of their topics, instead of being classified on the basis of their words. **Latent Dirichlet Allocation (LDA)** is the technique that is used to associate words with each topic, and to estimate which mixture of topics best describes each document.

Topic modelling applies data reduction techniques to reduce the dimensionality of the data. Previously, we discussed dimension reduction tools (in Chapter 7), where we specifically focused on PCA. These topics are low-dimensional factors

11.3 Topic Modelling

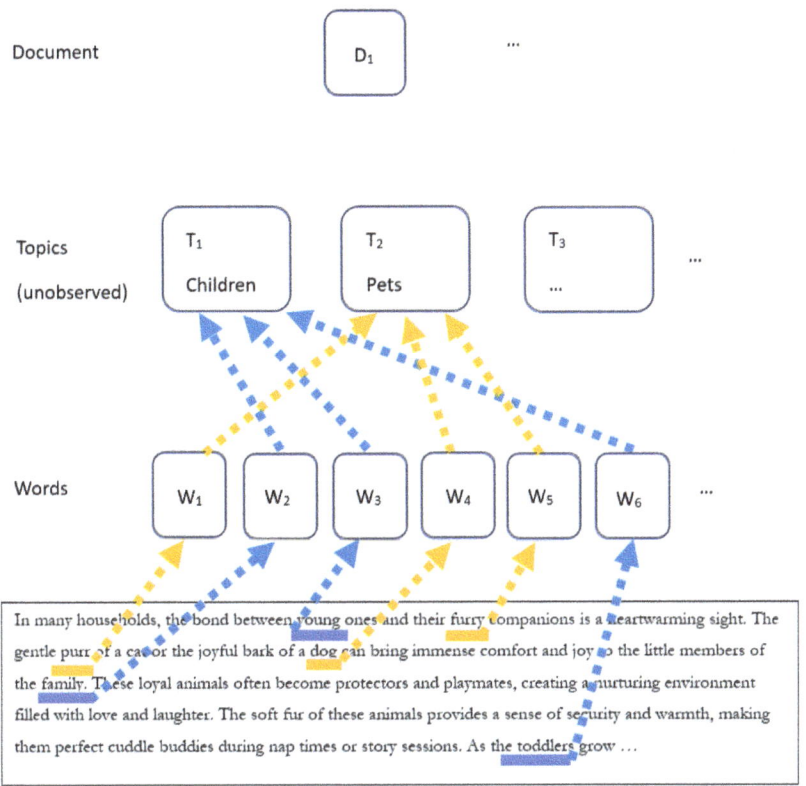

Fig. 11.8 Adding a topic layer (*Source* Author's elaboration, in the style of Provost and Fawcett 2013, Fig. 10.6)

that summarize high-dimensional data, in a similar way that PCA-generated variables are low-dimensional summary indicators that try to represent the larger set of input variables.

Topic modelling could potentially apply PCA (Grimmer et al. 2022), but PCA has some problems specifically in the area of text analysis,[5] and so a far more

---

[5] PCA in R normally uses the dense matrix format, which is problem for text analysis because the DTMs are sparse, and sparse matrices take up a lot of computational resources when they are stored with a dense matrix format. In contrast, LDA has a number of advantages, regarding being more interpretable, and more efficient in the sense of fewer factors to summarize a document. Topic model computation is computationally intensive in general, although there are many efficient LDA algorithms available. Furthermore, whereas PCA minimizes the sum of squares, LDA takes a different approach (based on multinomial deviance) which is more appropriate in the case of text analysis where many observations are zeroes (Taddy et al., pp. 330–331).

popular approach is Latent Dirichlet Allocation (LDA). LDA is an exploratory technique for the discovery of the unobserved "topics", and it is called "unsupervised" because the topics are not labelled prior to the analysis. Instead, the interpretations of the topics is a task for the researcher once the topics have been estimated. Like for PCA-generated variables, the topics that emerge from the data are not necessarily intelligible, but usually an interpretation or a label that is meaningful to humans can be given to these statistically-generated indicators.

Further Reading

Taddy et al (2023) contains a chapter on text analysis, with examples in R. Grimmer et al. (2022) is an in-depth book-length discussion of the theory of text analysis (although it does not have any empirical examples in R). The next two books are free to read online: Silge and Robinson (2017) on text analysis in R; with a follow-up book (Hvitfeldt and Silge 2022) that goes deeper into machine learning and text analysis in R.

References

Grimmer, J., Roberts, M. E., & Stewart, B. M. (2022). Text as data: A new framework for machine learning and the social sciences. Princeton University Press.

Hvitfeldt, E., & Silge, J. (2022). Supervised machine learning for text analysis in R. Chapman and Hall/CRC. https://doi.org/10.1201/9781003093459, Free to read online: https://smltar.com/

Mosteller, F., & Wallace, D. L. (1963). Inference in an authorship problem: A comparative study of discrimination methods applied to the authorship of the disputed Federalist Papers. Journal of the American Statistical Association, 58(302), 275–309.

Porter, M. F. (1980). An algorithm for suffix stripping. Program, 14(3), 130-137.

Provost, F., & Fawcett, T. (2013). Data Science for Business: What you need to know about data mining and data-analytic thinking. O'Reilly Media, Inc.

Silge, J., & Robinson, D. (2017). Text mining with R: A tidy approach. O'Reilly Media, Inc.

Taddy, M. (2019). Business data science: Combining machine learning and economics to optimize, automate, and accelerate business decisions. McGraw Hill Professional.

Taddy M., Hendrix L., Harding M.C. (2023). Modern Business Analytics. Practical Data Science for Decision Making. McGraw Hill, New York, NY.

Causal Inference   12

In many cases, we do not need causal knowledge because accurate predictions are sufficient. When deciding whether to bring an umbrella, you don't need to know how clouds work, or how umbrellas are made, or how weather processes can be manipulated; you merely need to know the prediction from a weather forecast (HBR 2023).

However, in many cases we require a knowledge of the underlying causal relations between variables. Consider the case of one-way train tickets and prices, in a context of price differentiation for commuters. In the early morning, there are many passengers, and prices are high. In the early afternoon, there are fewer passengers, and prices are lower. The statistical evidence would show that the number of tickets sold is positively correlated with prices. However, correlation does not imply causality. It would be wrong to conclude that the positive association (between number of tickets sold and prices) is evidence that prices have a positive causal effect on number of tickets sold. It would be disastrous to conclude that more tickets could be sold by doubling the prices. The true causal relationship should take into account the role of an unobserved confounding variable, which would be the high demand from commuters for morning travel. Even though we observe a positive correlation between prices and number of tickets sold, our background knowledge would lead us to expect a *negative* causal effect of prices on number of tickets sold. The correct causal interpretation can sometimes be the opposite of the observed correlation. Furthermore, notice that some business decisions (such

---

Supplementary Information The online version contains supplementary material available at https://doi.org/10.1007/978-981-95-2433-4_12.

as deciding how to set prices) require causal knowledge rather than knowledge of associations.

This chapter discusses the importance of causal understanding, and how causal knowledge refers to a higher level of understanding than what can be observed from correlations in data. Statistical techniques designed for making causal interpretations are also discussed, such as Randomized Controlled Trials (RCTs) and pseudo-experimental methods.

12.1 Correlation Is Not Causation

It is often stated that correlation is not the same as causation, although in practice confusion remains. If two variables are correlated, it is often casually suggested that one variable is influencing the other. However, a correlation between two variables could mean as many as five different things, as shown in Fig. 12.1.

Figure 12.1 highlights how an observed correlation between two variables X and Y can be given as many as 5 different causal interpretations. These are discussed in turn:

a. $X \rightarrow Y$: The correlation between X and Y corresponds to a causal effect of X on Y
b. $X \leftarrow Y$: Here, the correlation between X and Y indicates a causal effect of Y on X
c. $X \leftarrow Z \rightarrow Y$: Z acts as a confounding variable. X and Y are correlated, but we should not interpret this as a causal effect of one variable on the other. This is a common case in social science research, where background variables affect both X and Y.
d. $X \rightarrow$ collider $\leftarrow Y$: This is known as collider bias (Elwert and Winship 2014). X and Y seem to be correlated in our data sample, but X and Y would be uncorrelated in the overall population. The correlation between X and Y arises because of selection bias. A first example would be the case of a sample of patients in hospital, whose likelihood of disease X might be correlated with the likelihood of disease Y, simply because having (at least) one disease increases one's chances of being admitted to hospital. A correlation between diseases X and Y might be observed in the (unrepresentative) sample of hospital patients; even if diseases X and Y are unrelated in the overall population (Pearl and MacKenzie 2018, p. 197). A second example would be that, even if physical beauty and acting talent are statistically unrelated in the overall population, a focus on a subsample of Hollywood actresses (who get hired either on account of their talent, or on account of their beauty) might reveal a negative correlation, such that exceptional beauty could indicate mediocre acting skills (Cunningham 2021).
e. Finally, X and Y might be correlated, to a statistically significant degree, simply because of chance. Such chance correlations are of course unlikely to repeat themselves in new data samples. The appearance of correlations that are simply

12.1 Correlation Is Not Causation

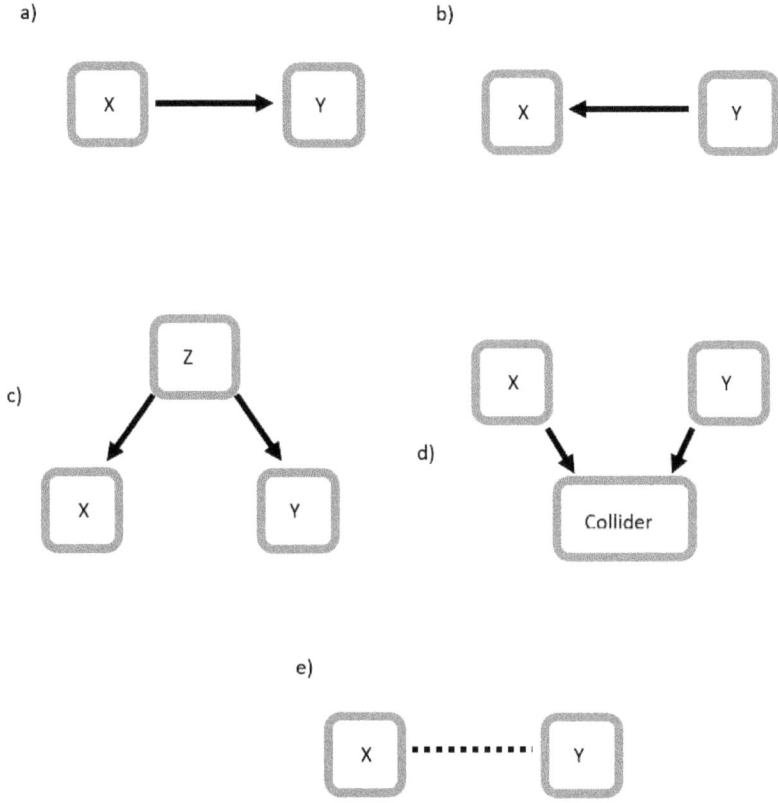

Fig. 12.1 If X and Y are correlated, this could mean five things (*Notes* Author's elaboration. *Arrows* denote the direction of causality)

due to chance is a potential problem in data mining applications, where analysts search for correlations in large datasets in the hope of discovering new patterns in the data (Peters et al. 2017). Hopefully, the role of such chance correlations between variables will be removed by using procedures designed to handle overfitting, such as cross-validation.

Figure 12.1 highlights how the observation that X and Y are correlated does not, in itself, give us causal understanding about the relationship between X and Y. There is a big leap that is required to move from the level of observed correlations, to the level of causal understanding. Data per se "are profoundly dumb" (Pearl and MacKenzie 2018, p. 6); "Data … are totally oblivious to cause-effect asymmetries" (Pearl and MacKenzie 2018, p. 99); and "you cannot draw causal conclusions without some causal hypotheses" (Pearl and MacKenzie 2018, p. 79).

Valuable causal knowledge does not come from the data per se, but from our (causal) interpretations of the data:

> Two people who believe in two different causal diagrams can analyze the same data and may never come to the same conclusion, regardless of how "big" the data are. (Pearl and MacKenzie 2018, p. 90)

Collecting more data will not solve the problem of how to give causal interpretations to observed correlations. In the age of abundant data, knowledge of the causal relations between variables is valuable and rare.

12.2 Causal Language

Given the many possible causal interpretations that can belie an observed correlation, it is important to be careful when presenting your results, and to be skeptical when others present their results in causal terms. It often happens that people inadvertently suggest causal relations between variables, where such causal interpretations are not warranted on the basis of the evidence. If you are a manager discussing with your "quants", and they use words such as "driver of" and "effect on", they are giving causal interpretations to what may simply be observed correlations, and it is better to check with them whether their suggestions of causality are truly warranted, or if other causal interpretations are possible. Table 12.1 distinguishes between the language of associations, and the language of causality.

A helpful definition of causality is the **"do" operator** of Pearl (2009). Consider first the case of a positive correlation between X and Y: when Y is above average, X is also expected to be above average. This knowledge of how X and Y co-vary does not give us useful knowledge regarding how Y will change if we vary X. The do operator focuses explicitly on interventions, asking questions such as: What is the expected value of Y if we fix X to take the specific value x? This would be denoted as $P(Y|do(X = x))$. For example, what is the expected value of Y if we intervene to change X from having an above-average value to fix it at a below-average value? Will our manipulation of variable X lead to any change in Y? If fixing X at a new value x does not lead to any change in Y, then X has no causal effect on Y. This knowledge of the causal effect of X on Y comes from intervening on X, it does not come from merely observing X.

Table 12.1 Language of associations vs language of causality

Associations	Causal effects
Associated with…	Driver of…
Correlated with…	Effect on…
Linked to…	Impact on…
Related to…	Influences…

12.3 Directed Acyclic Graphs (DAGs)

Pearl (2009) gives the example of a barometer. We observe that the movement of the barometer needle is associated with changes in the probability of rain. Superstitious learning might lead some to believe that the movement of the barometer needle causes the rain. However, such a causal interpretation could be refuted by applying the reasoning of the do operator to the case of the barometer needle. If we were to intervene by manually forcing the barometer needle to give a low pressure reading, this would have no effect on the probability of rain. By intervening on one variable, and observing whether another variable changes, allows us to transcend knowledge of correlations to reach the level of causal knowledge.

Another example would be case c) in Fig. 12.1. X and Y are correlated, but the relationship between them is driven by a **confounding variable**: $X \leftarrow Z \rightarrow Y$. If we could intervene on X to fix it at a value $X = x$, denoted as "do($X = x$)", this manipulation would not affect the outcomes for Y, because there is no causal path from X to Y. If the confounding variable Z is observed, then we can adjust for it (e.g. adding Z as a control variable in a regression) to show that X and Y are unrelated once the influence of Z is removed. In many cases, though, the confounding variable Z is unobserved.

12.3 Directed Acyclic Graphs (DAGs)

Directed Acyclic Graphs (DAGs) are a simple yet powerful way for thinking about the causal structure between variables. DAGs are shown in Fig. 12.1, with nodes (corresponding to variables) and directed arrows (corresponding to the direction of causal inference). DAGs help to resolve the ambiguity in causal reasoning that can come from the use of mathematical equations (Pearl 2009; Coad 2021).

DAGs can clarify whether or not you should include a control variable. A first example relates to drink driving. In the R code file, we analyse a (fictional) dataset (drink.xlsx) with three variables: number of beers drunk in the previous 3 hours, blood alcohol level, and driving ability. Table 12.2 reports the regression results. Model 1 shows a clear negative association between drinking beers and driving ability. Model 2 adds a further control variable (blood alcohol level) and now finds that there is no statistically significant relationship between drinking beers and driving ability. Model 2 is clearly a better fitting model, because the adjusted R2 statistic is higher (0.98, compared to 0.92 for Model 1).

Should we prefer the results from Model 2?
Can we conclude that it is safe to drink beers and go driving?
If not, why not?

A DAG representation helps to clarify the situation. Presumably the significant association in Model 1 corresponds to a causal effect of beer drinking on driving ability. Model 2 presumably corresponds to the following DAG: Beer drinking $\rightarrow$ Blood alcohol level $\rightarrow$ Driving ability. If we want to know whether beer

Table 12.2 Beer drinking and driving ability example

	Model 1		Model 2	
Intercept	71.18	***	71.46	***
Std error	0.74		0.41	
Beers	−8.47	***	0.17	
Std error	0.35		0.83	
Blood alcohol			−8.57	***
Std error			0.80	
Adjusted R2	0.92		0.98	

Notes N = 50 observations
*** Denotes statistical significance at the 1% level. See R code file for details

drinking is related to driving ability, we should not control for blood alcohol level, which lies on the causal path from beer drinking to driving ability.[1]

This example has shown us the problems that arise from the common practice of adding as many control variables as you can find in your dataset. Adding control variables can sometimes lead to incorrect conclusions. Furthermore, if regression models (such as Model 2) show that a variable's coefficient is not statistically significant, this does not mean that there is no causal relation. In this case, the problem arose because the blood alcohol variable causally mediates the influence of beer drinking on driving ability. Managers should be cautious when interpreting the analytical results of others, recognizing that regression models with more explanatory variables are not necessarily better, and that a DAG representation can help to decide whether explanatory variables should be included or excluded.

The value of DAGs for decision-making is illustrated in another example. Consider Table 12.3, which refers to results of a clinical trial of a new drug. For females, the heart attack rate is 10% in the control group, and 15% in the treatment group. For males, the heart attack rate is 25% in the control group, and 30% in the treatment group. So far, it seems like the outcomes for the treatment group are worse than the outcomes for the control group. However, focusing on the full sample, the heart attack rate is lower in the treatment group (17%, compared to 20%). This is a life or death situation, and it is important to make the right decision.

Should we conclude that the treatment is good or bad?
If we are interested in the overall effectiveness of the treatment, should we control for gender or not?

---

[1] This is known as the "bad controls" problem (Angrist and Pischke 2008). However, if we know that beer drinking affects driving ability, and are interested in the mechanisms through which beer drinking affects driving ability, then adding blood alcohol levels as an explanatory variable could be a useful approach to decompose the mechanisms of the effect of beer drinking on driving ability.

Table 12.3 Example of Simpson's paradox: medical treatment and heart attacks

	Control		Treatment	
	Heart attack	No heart attack	Heart attack	No heart attack
Female	2	18	6	34
Male	10	30	6	14
Total	**12**	**48**	**10**	**50**
Heart attack rates				
Women (%)	10		15	
Men (%)	25		30	
Total (%)	20		17	

Notes Author's elaboration, inspired by Pearl and MacKenzie (2018, p. 201)

This example is an instance of **Simpson's Paradox**. Controlling for gender leads to a reversal of the main findings. The answer becomes clearer once we reformulate the problem in DAG notation (see Fig. 12.2).

Figure 12.2 reminds us of what we are interested in: the causal effect of treatment on the outcome. Gender has an effect on treatment status (40/60 females are in the treatment group, compared to only 20/60 males). Gender also has an effect on the heart attack rate (8/60 heart attacks for females, compared to 16/60 heart attacks for males). However, to recover the causal effect of the treatment on heart attacks, we need to remove the confounding role of gender. Therefore, we decide to hold gender constant, and separately examine the results for the subset of females (the treatment is bad) and the subset of males (the treatment is bad). We conclude that the treatment is bad. Note how a DAG representation gives us fairly clear guidance to give a normative answer to the question of whether or not control variables should be included.

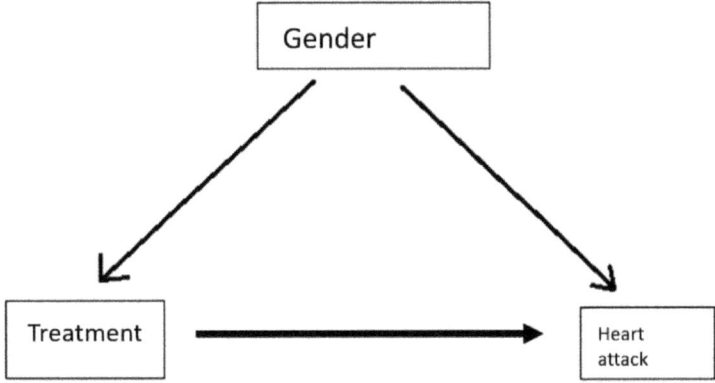

Fig. 12.2 DAG representation of the example in Table 12.3

12.4 Techniques for Causal Inference

Econometricians have long thought about how it might be possible to obtain causal understanding from the mere observation of empirical evidence. To this end, a number of techniques have been proposed, that are studied in this section. An important distinction is made between experimental data and observational data. With **experimental data**, the researcher can manipulate the treatment assignment (e.g. using randomization) to make sure that self-selection bias has no role (i.e. participants cannot self-select into their preferred group on the basis of their characteristics). With **observational data**, however, outcomes are observed for various participants, but we cannot be sure whether these participants self-selected into their groups on the basis of initial differences. A major problem for causal inference is **self-selection**: whether individuals can choose which group they go into. For example, if individuals can choose to be smokers or non-smokers (e.g. on the basis of genetic factors), then any difference in outcomes could be due to genetic differences at the start.

12.4.1 Randomized Controlled Trials (RCTs)

Randomized Controlled Trials (RCTs) are widely regarded as the "gold standard" for causal inference. With RCTs, the researcher can avoid the problem of confounders (Z in case (c) of Fig. 12.1) and thereby interpret a correlation as a causal effect. Figure 12.3 illustrates how RCTs allow for causal interpretations. Figure 12.3 left shows that, in general, a correlation between the treatment and the outcome cannot be given a causal interpretation, because the correlation between Treatment and Outcome mixes together two effects: first, the direct causal effect that is represented by the arrow Treatment → Outcome (case (a) in Fig. 12.1); and second, the indirect effect represented by the "**backdoor path**" which is the case of Treatment ← Confounder → Outcome (case (c) in Fig. 12.1). If membership of the Treatment group is assigned at random, however, then Treatment is no longer systematically related to the Confounder, and the link from Confounder to Treatment is cut (Fig. 12.3, right). In this case, the correlation between Treatment and Outcome can be given a causal interpretation.

RCTs thus enable researchers to estimate the causal effect of a treatment on an outcome. As a result, RCTs have been met with much enthusiasm by economists: "Randomized experiments do occupy a special place in the hierarchy of evidence, namely at the very top" (Imbens 2010, p. 407).

12.4.2 Natural Experiment

A natural experiment aspires to be like an RCT, even though it is based on observational data rather than experimental data. In terms of Fig. 12.3 (right), the link from the confounding variable to treatment assignment is cut, because treatment

12.4 Techniques for Causal Inference

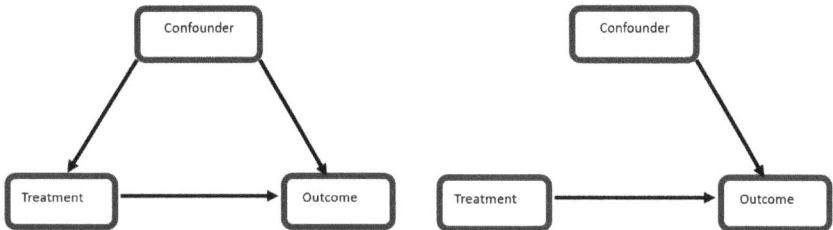

Fig. 12.3 *Left* The correlation between treatment and outcome cannot be given a causal interpretation, because of the existence of a confounder. *Right* Randomization cuts the arrow from confounder to the treatment; therefore any correlation between treatment and outcome can be given a causal interpretation Treatment → Outcome (*Source* Author's elaboration)

assignment is due to some strong external factor. In other words, the randomization is forced by some "natural accident." A well-known example of a natural experiment is the case of the 1854 Broad Street cholera outbreak in London. At the time, it was thought that cholera was transmitted through the air, but now it is known that cholera is transmitted through contaminated water. Half of the neighbourhood was drinking contaminated water delivered by one water company (water obtained downstream from a sewage discharge), while the other half was drinking cleaner water delivered by a different water company. Infection rates closely followed the haphazard contours of the different service areas of these two companies, even if the reasons behind the differences in outcomes were not initially understood. The physician John Snow, who identified the water supply as the source of the outbreak, described the situation as "an experiment… on the grandest scale."[2]

12.4.3 Regression Discontinuity Design (RDD)

Consider the case of a competition for a research and development (R&D) grant, where 100 firms apply but only the best 10 applicants will succeed. A committee of experts ranks the applicants, and the 10 highest-scoring applications will receive the grant. In this situation, the #1 top-ranking firm can be expected to be very different from the #100 lowest-ranking firm. However, the differences between the characteristics of the firms ranking as #10 and #11 are likely to be negligible at the time when the grants are distributed. If the firm ranking as #10 embarks on a period of superior performance in the subsequent time period (when compared to the #11 firm), this cannot be attributed to any big differences between the firms at the time of grant award, but would instead be interpreted as a causal effect of receiving the grant.

---

[2] See https://en.wikipedia.org/wiki/Natural_experiment (last accessed 26th July 2025).

Such is the intuition of Regression Discontinuity Design (RDD). Observations are sorted in terms of a "**running variable**" or "forcing variable", and the causal effect of a treatment can be identified in the local vicinity of the threshold. A strength of the RDD approach is that it offers a credible strategy for causal inference. A number of drawbacks can be mentioned, however. First, the data has to be organized in a very specific way, in terms of the running variable. Second, it is not possible to extrapolate from the results obtained close to the threshold, to try to generalize about results for all of the observations. In the context of the example given above, RDD helps us to estimate the causal effect of the treatment for the firm ranked as #10, but we cannot estimate the treatment effect of the R&D grant for the firm ranked as #1.

12.4.4 Instrumental Variables (IV)

Consider the case in Fig. 12.4 of two variables, x and y. The correlation between x and y cannot be interpreted as a causal effect of x on y, because there is the confounding influence of an unobserved variable u. It might be possible to obtain data on a suitable variable z, that is satisfies the following two requirements of an instrumental variable (IV):

- The IV "**first stage**" condition: There should be a significant correlation between z and x that corresponds to a direct causal influence of z on x.
- The **Exclusion Restriction**: There should be a significant correlation between z and y, that corresponds to a direct causal influence of z on y that is fully mediated by the causal pathway that operates via x. Controlling for x, there is no relationship (neither a correlation nor a causal relationship) between z and y.

These two pieces of causal knowledge allow us to calculate a third: the causal effect of x on y (Pearl 2009; Coad 2021). IV is not a way of creating causal knowledge on the basis of correlations; instead IV can only be expected to create causal knowledge if you bring your own causal knowledge to begin with. If your choice

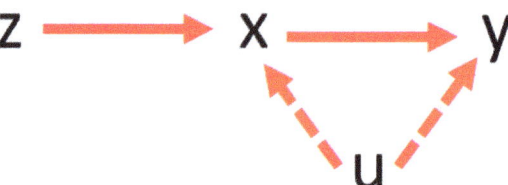

Fig. 12.4 DAG for understanding identification in Instrumental Variables (IV) analysis

of z is based on questionable causal assumptions regarding the two conditions mentioned above, then you should not expect much from your IV estimates.

A major drawback of these two conditions is that they are fundamentally untestable: this is because observational data can tell us if z and x are correlated, but observational data cannot tell us whether z causes x. You just have to believe. No doubt, there have been many abuses of IV in the past, and we might never know.

Further Reading

Pearl and MacKenzie (2018) is a reader-friendly discussion of the theory of causality. Huntington-Klein (2021) discusses research design and causal inference. Several books discuss causal inference by also including examples of working with data in R: Cunningham (2021), Imai and Williams (2022) and Taddy et al. (2023). Cunningham (2021) is free to read online.

References

Angrist, J. D., & Pischke, J.-S. (2008). Mostly harmless econometrics: An empiricist's companion. Princeton University Press.

Coad, A. (2021). Econometrics and the growth of firms: perspectives from evolutionary economics. Strategy Science, 6 (4), 338–352. https://doi.org/10.1287/stsc.2021.0132

Cunningham, S. (2021). Causal inference: the mixtape. Yale University Press: New Haven, CT, USA. Free to read online: https://mixtape.scunning.com/index.html

Elwert, F., & Winship, C. (2014). Endogenous selection bias: The problem of conditioning on a collider variable. Annual Review of Sociology, 40, 31–53.

HBR (2023). HBR guide to AI Basics for Managers. Harvard Business Review Press: Massachusetts, USA.

Huntington-Klein, N. (2021). The Effect: An Introduction to Research Design and Causality. Chapman and Hall/CRC.

Imai, K., & Williams, N. W. (2022). Quantitative Social Science: An Introduction in Tidyverse. Princeton University Press.

Imbens G. W. (2010). Better LATE than nothing: some comments on Deaton 2009 and Heckman and Urzua 2009. Journal of Economic Literature 48, 399–423.

Pearl, J., & Mackenzie, D. (2018). The Book of Why: The New Science of Cause and Effect. Basic Books: USA.

Pearl, J. (2009). Causality: Models, reasoning and inference. 2nd Edition. Cambridge University Press: Cambridge, UK.

Peters, J., Janzing, D., Scholkopf, B. (2017). Elements of Causal Inference: Foundations and Learning Algorithms. MIT Press: Cambridge, MA.

Taddy, M., Hendrix, L., & Harding, M. C. (2023). Modern Business Analytics. Practical Data Science for Decision Making. McGraw Hill: New York, NY.

Concluding Remarks 13

This book provided an overview of many theoretical perspectives on digital transformation, as well as quantitative techniques for data analysis. The book has hopefully offered an interesting and useful overview of the area, although there are limits in terms of the number of topics covered, as well as the depth of treatment of topics. This is a fast-changing area, and there are many emerging developments, such as neural networks, deep learning, and new AI techniques. It is therefore up to the reader to continue the journey, and to be continually investing in learning and updating their knowledge in this area.

Managers do not need to know all the details of advanced quantitative analytics and AI methods, but it is important to have a solid foundation, and to ask analysts the right questions, such as: Are these results merely associations, or can we give a causal interpretation? What alternative explanations might explain the results? Have these results been validated on test data? What are the assumptions underlying the techniques used? What are the limitations of the analysis? What are the main concerns about data quality? What are the ethical nightmare situations we should consider? And so on.

A final idea is that data scientists should be able to explain things in a language that is widely understand. If, after reading this book, you still do not understand what a data scientist (or consultant presenting a quantitative analytics proposal) is talking about: maybe the problem is not you, it could be them!

Bibliography

Chapman, A., Simperl, E., Koesten, L., Konstantinidis, G., Ibáñez, L. D., Kacprzak, E., & Groth, P. (2020). Dataset search: a survey. The VLDB Journal, 29(1), 251–272.
Davenport, T. H., & Miller, S. M. (2022). Working with AI: real stories of human-machine collaboration. MIT Press: Cambridge, MA.
Davenport, T. H., & Mittal, N. (2023). All-in on AI: How smart companies win big with artificial intelligence. Harvard Business Press: Cambridge, MA.
Halevy, A., Korn, F., Noy, N. F., Olston, C., Polyzotis, N., Roy, S., & Whang, S. E. (2016). Goods: Organizing google's datasets. In Proceedings of the 2016 International Conference on Management of Data (pp. 795–806). Crossref DOI link: https://doi.org/10.1145/2882903.2903730.
Hastie, T., Tibshirani, R., Friedman, J. H. (2017). The Elements of Statistical Learning: Data Mining, Inference, and Prediction. Second Edition. New York: Springer.
Kelleher, J. D. (2019). Deep learning. MIT Press: Cambridge, MA
Kleinberg, J., Ludwig, J., Mullainathan, S., & Obermeyer, Z. (2015). Prediction policy problems. American Economic Review, 105(5), 491–495.
Lee, K. F. (2018). AI superpowers: China, Silicon Valley, and the new world order. Houghton Mifflin.
Miloslavskaya, N., & Tolstoy, A. (2016). Big data, fast data and data lake concepts. Procedia Computer Science, 88, 300–305.
Nagelkerke N. J. D. (1991). A note on a general definition of the coefficient of determination. Biometrika 78 (3), 691–692.

Index

A
Accountability, 75, 86
Agile teams, 32, 37, 38
AICc, 67, 69
AI ethics, 75, 77, 86, 87
AI Ethics Committee, 87
Analytics sandbox, 38, 39
Application Programming Interfaces (APIs), 35–37, 39–41
Applied Statistics, 8, 43, 92, 96
Artificial Intelligence (AI), 1, 2, 4, 6–10, 12, 21, 24–27, 32, 43, 46–51, 74–88, 91, 125, 165
Associations, 5, 8, 18, 46, 88, 153, 154, 156, 157, 165
Automatic Interaction Detection (AID), 130

B
Backdoor path, 160
Bag of words, 140, 143, 146, 147, 149
Big data, 1, 2, 4, 6–10, 33, 41, 44, 46, 47, 63, 65, 66, 105, 129
Big Data Business Model Maturity Index (BDBMMI), 21–24
Binned scatterplot, 120, 121
Black box algorithms, 79
Business Intelligence (BI), 21, 44

C
Causal inference, 44, 139, 157, 160, 162, 163
Chartjunk, 114
Classification and Regression Trees (CART), 125–130, 132–134, 137
Cloud, 6, 18, 20, 33–35, 39, 41, 80, 101, 105–107, 120, 145, 153
Confounding variable, 153, 154, 157, 160

Confusion matrix, 83, 85
Contour plot, 120
Convergent experimentation, 21
Corpus, 141, 142, 144, 145, 147
Correlations, 54, 55, 63, 71, 119, 134, 153–157, 160–162

D
Dashboards, 23, 38, 110, 111, 113, 117, 122, 123
Data imputation, 96, 97
Data-ink ratio, 114
Data lake, 28, 32–34, 38, 39, 106
  FAIR principles, 34
Data monetization, 21, 23, 24, 27
Data pre-processing, 95
Data quality, 34, 45, 81, 91–93, 165
Data Science, 2, 7–9, 23, 26, 33, 38, 39, 43, 44, 46, 48, 49, 54, 64, 74, 87, 88, 92, 96, 99, 121, 123, 139
Data visualization, 45, 46, 95, 96, 110, 114, 117, 118, 122, 123
Data warehouse, 7, 20–22, 33, 34
Decision trees, 127, 128
Dependent variable, 55, 56, 59–61, 63, 68, 70, 71, 80, 87, 88, 96, 133, 135, 136, 147
Deviance, 59, 151
Digitalization, 1, 2, 5, 6, 11, 75
Digital Transformation (DX), 1, 2, 5–7, 10, 11, 17, 18, 20, 21, 24, 26–28, 34, 35, 38, 41, 46–51, 105, 107–109, 165
Digitization, 1, 2, 5, 6
Dimensionality reduction, 95
Directed Acyclic Graphs (DAGs), 157–159, 162
Discretization, 95, 96

Divergent experimentation, 21
Document Term Matrix (DTM), 140, 143, 144, 147, 149, 151
"Do"operator, 156

E
Electrification, 10, 11
Ethics, 74, 75, 81, 87, 88, 94
Exclusion Restriction, 162
Experimental data, 160
Explainability, 76, 79, 80, 86
Explanatory variable, 56, 59, 61, 64–66, 68, 69, 71–73, 88, 96, 126, 127, 129, 133–136, 141, 158

F
Factory system, 31
Fairness, 82–84, 86
False negative, 63, 83–85
False positive, 63, 83–85
Features, 23, 27, 31, 50, 64, 68, 108, 115, 117, 127, 133, 136
Fusion skills, 48

G
Gamlr, 67–70, 134
Garbage In, Garbage Out (GIGO), 92
General Purpose Technology (GPT), 11, 12
Greedy algorithms, 128

H
High-dimensional datasets, 64
Highest Paid Person's Opinion (HIPPO), 49
Histogram, 53, 54, 66, 67
Human in the loop for exceptions (HITLFE), 85
Human in the loop (HITL), 85, 86
Human on the loop (HOTL), 85
Human out of the loop (HOOTL), 85, 86

I
In-sample, 64
Instrumental Variables (IV), 162, 163
Intercept term, 56, 58, 60, 64, 66, 68, 69
Inverse Hyperbolic Sine (IHS), 68, 96, 101

J
Jitter, 119, 120
Judgment, 76, 77

K
Kernel density, 53, 54

L
Lasso regression, 63, 66, 70, 141, 146
Latent Dirichlet Allocation (LDA), 150–152
Logarithmic transformation, 54, 96, 100, 101
Logistic regression, 59–63, 70, 71, 80, 87, 88, 125
Log odds, 59, 60

M
Machine learning, 2, 6–9, 26, 41, 44, 46, 59, 75, 78–80, 83, 85, 91, 152
  supervised, 9
  unsupervised, 9, 101, 152
Marketing funnel, 18, 19
Microservices, 35–37
Minimum Viable Prototype (MVP), 20, 37, 38, 108
Missing values, 67, 68, 92, 93, 96–98, 100, 101, 133, 134
Modular IT, 37
Monolithic IT, 36, 37
Multidivisional corporation, 31, 32

N
Nagelkerke R2 statistic, 59, 62, 87
Natural Experiment, 160, 161

O
Observational data, 160, 163
OLS Linear Probability Model (OLS-LPM), 63
OLS linear regression, 55, 60, 62–66, 71, 80, 125, 126, 130
Organizational Culture, 26, 49, 51
Organizational Structure, 1, 26, 32, 108
Outliers, 81, 94, 96, 97, 113, 119
Out-of-sample, 64, 132
Overfitting, 64–66, 119, 120, 125, 126, 132, 136, 146, 155
Overplotting, 120

Index

P
Pie charts, 117, 118, 122
Point solutions, 11–13
Pre-attentive processing, 115, 122
Predictions, 8, 22, 38, 56, 58, 61, 63, 65, 76–78, 80–82, 85, 110, 126, 127, 129, 130, 132, 139, 153
Prescriptions, 22
Principal Components Analysis (PCA), 95, 98–102, 150–152
Principles of Visual Design, 116
Privacy, 27, 76, 77, 81, 82, 120
Productivity J-curve, 12
P-value, 56, 60, 62

R
R2 statistic, 157
Random Forests, 80, 125, 126, 132, 133, 136, 137
Randomized Controlled Trials (RCTs), 154, 160
Regression, 43, 54, 56–74, 80, 87, 88, 94–97, 119, 128–130, 141, 146, 157, 158
Regression Discontinuity Design (RDD), 162
Rule of thirds, 116, 117

S
Sampling, 95
Scatterplot, 55, 73, 94, 96, 114, 119–121, 134
Scatterplot matrix, 55, 99–102
Self-selection, 160
Sentiment analysis, 41, 147, 150
Sentiment lexicon, 147–149
Signal and noise, 63
Siloes, 25, 26, 32, 36
Simpson's paradox, 159
Single Source of the Truth (SSOT), 33, 39
Skewed data, 54, 94
Skills of a data scientist, 44
Skills of leaders, 46
Stemming, 143
Stopping rule, 128
Stopwords, 141–145
System solutions, 11, 13

T
Technical debt, 40
Term frequency, 145
Test data, 64, 165
Text as data, 139, 142
Text regression, 141, 146, 147
Tf-idf, 145, 146
3D plots, 118
Tokenization, 140–145
Topic modelling, 149–151
Training data, 13, 64, 79, 83
Trees. *See* CART
T statistic, 56

U
User Experience (UX), 105, 106, 108–111

V
Variable transformation, 95, 96

Z
Z pattern of scanning, 116, 117

GPSR Compliance
The European Union's (EU) General Product Safety Regulation (GPSR) is a set of rules that requires consumer products to be safe and our obligations to ensure this.

If you have any concerns about our products, you can contact us on

ProductSafety@springernature.com

In case Publisher is established outside the EU, the EU authorized representative is:

Springer Nature Customer Service Center GmbH
Europaplatz 3
69115 Heidelberg, Germany

www.ingramcontent.com/pod-product-compliance
Ingram Content Group UK Ltd.
Pitfield, Milton Keynes, MK11 3LW, UK
UKHW022203230426
470311UK00001BA/14